SECURING CYBERSPACE

A New Domain for National Security

FOREWORD BY

JOSEPH S. NYE & BRENT SCOWCROFT

EDITED BY

NICHOLAS BURNS & JONATHON PRICE

The Aspen Institute
One Dupont Circle, N.W.
Suite 700
Washington, DC 20036

Published in the United States of America in 2012 by The Aspen Institute

Cover Design by: Steve Johnson
Interior Layout by: Sogand Sepassi

aspen strategy group

CO-CHAIRMEN

Joseph S. Nye, Jr.
University Distinguished Service Professor
John F. Kennedy School of Government
Harvard University

Brent Scowcroft
President
The Scowcroft Group, Inc.

DIRECTOR

Nicholas Burns
Professor of the Practice of Diplomacy
and International Politics
John F. Kennedy School of Government
Harvard University

DEPUTY DIRECTOR

Jonathon Price
Deputy Director
Aspen Strategy Group

ASSOCIATE DIRECTOR

Jennifer Jun
Associate Director
Aspen Strategy Group

MEMBERS

Madeleine Albright
Chair
Albright Stonebridge Group

Graham Allison
Director, Belfer Center
John F. Kennedy School of Government
Harvard University

Zoë Baird Budinger
President
Markle Foundation

Samuel R. Berger
Chair
Albright Stonebridge Group

Stephen E. Biegun
Vice President
Ford Motor Company

Robert D. Blackwill
Henry A. Kissinger Senior Fellow for
U.S. Foreign Policy
Council on Foreign Relations

Eliot Cohen
Professor
Johns Hopkins SAIS

Susan Collins
Senator
United States Senate

Richard Cooper
Professor
Harvard University

Richard Danzig
Chairman
Center for a New American Security

John Deutch
Institute Professor
Massachusetts Institute of Technology

Acknowledgements

Nicholas Burns
Director, Aspen Strategy Group

Jonathon Price
Deputy Director, Aspen Strategy Group

In early August 2011, the Aspen Strategy Group convened for a week in Aspen, Colorado to examine a novel and formidable national security challenge: cybersecurity. The diverse group of ASG members and invited guests – comprising government officials, policymakers, academics, journalists, corporate leaders, and foreign policy experts – ensured a wide range of perspectives and a bipartisan dialogue about the complex challenges interwoven in cyberspace. This publication presents the policy papers that established the foundation for our discussions throughout the week in Aspen. It also includes the third annual Ernest May lecture delivered by ASG co-chair, Joseph Nye.

As always, the Aspen Strategy Group would like to express our sincerest gratitude to a number of organizations and individuals whose dedication, generosity, and support make the summer workshop possible. Our supporters for this year's workshop include Mr. Howard Cox, the Bill & Melinda Gates Foundation, the Markle Foundation, McKinsey & Company, Mr. Simon Pinniger, the Margot & Thomas Pritzker Family Foundation, the Resnick Family Foundation, Ms. Carolyne Roehm, the Stanton Foundation, and Mrs. Leah and Mr. Ralph Wanger.

We would also like to thank our Associate Director Jennifer Jun and our Brent Scowcroft Award Fellows Allie Kirchner and Annie Moulton for their important contributions to this initiative. We look forward to following their careers as the next generation of foreign policy leaders and experts. Rebecca Yael Weissburg provided her invaluable proofreading and editing skills, and we are deeply grateful.

Finally, our highest regards go to our co-chairmen, Joseph Nye and Brent Scowcroft, whose vision and expertise provide the founding merits of the Aspen Strategy Group. Only through their leadership has the ASG been able to build a legacy as a forum of innovative strategic thinking on the most complex national security issues of the day.

Contents

Part 1

CYBERWARS & CYBERTERROR:
UNDERSTANDING CYBERSPACE AS A NEW BATTLEGROUND

Part 2

CYBER POLICY: REGULATING CYBERSPACE

Part 3

CYBERCRIME: IMPLICATIONS FOR BUSINESS AND THE ECONOMY

Part 4

CYBERSECURITY AND ITS TENSIONS WITH INTERNET FREEDOM

Part 5

CYBERSPACE: NEW POLICIES AND A NEW STRATEGY

Foreword
by ASG Co-Chairmen

Joseph S. Nye, Jr.
ASG Co-Chairman

Brent Scowcroft
ASG Co-Chairman

Last March, the U.S. Department of Defense reported that in one of the largest cyberattacks in U.S. history, unidentified hackers targeted the Pentagon, stealing approximately 24,000 files, some of which contained highly sensitive information on aircraft avionics, surveillance technologies, satellite communications systems, and network security protocols. Although the scope and success of the attack garnered a great deal of public attention, it is simply one of the countless attacks targeting the government, corporations, and private citizens that rely on cyber technology to meet their daily and perennial needs in areas of communication, business and finance, infrastructure, transportation, and much more. The indispensable role that cyber has come to play in virtually all areas of life renders us all the more vulnerable to the consequences of an insecure cyber environment.

Cyberspace today is marred by new and difficult challenges that we have not yet experienced in other traditional domains such as land, sea, air, and space. In the physical realm, countries are marked by borders, and individuals can be traced by addresses, phone numbers, and identification cards. In cyberspace, national borders and individual identities are more ambiguous, making it difficult to attribute a cyberattack to a specific actor. Another perplexing challenge is the inadequate understanding of the full dangers associated with cyberattacks. The private sector's natural sensitivity to investor confidence proves a major disincentive for companies to reveal any information about their vulnerability and experiences of attack. Meanwhile, governments are unlikely—and rightly so—to publicize information that may compromise national security, including past cases of attacks orchestrated by foreign entities. As for private citizens, many are victims or even abettors of widespread attacks, all without their knowledge.

As a group of national security experts that seek to apply our collective acumen to tackle the most important challenges to the United States and the wider world, we chose to convene our annual Summer Workshop around the topic of cybersecurity. Our aim was to gather around one table the most knowledgeable and trusted experts on cyberspace to share knowledge, build understanding, and discuss possible policy responses for the current and future challenges of cyber. The result was rewarding. Enriched by the participation of key government experts on cyber including Deputy Secretary of Defense William Lynn, former Vice Chairman of the Joint Chiefs of Staff General James Cartwright, USCYBERCOM Commander General Keith Alexander, and Deputy Secretary of Homeland Security Jane Holl Lute, and others, the five-day workshop examined the concepts of cyber war, terror, and crime; discussed offensive capabilities and defensive responses in this new medium; explored both the national and international frameworks that seek to regulate cyberspace; identified the threat posed to the private sector and critical infrastructure; and studied the relationship between cybersecurity and internet freedom. We also debated the gravity, scope, and immediacy of the threat—and ultimately concluded that these dangers pose grave problems for the United States that demand immediate attention and action. After five days of discussion, some of our main conclusions to emerge from the workshop include the need to increase public awareness of the cyber threat, the desirability to develop international norms for cyberspace, the importance of building private-public partnerships to secure cyberspace, and the continued role for U.S. leadership on the international stage.

The Aspen Strategy Group, a policy program of the Aspen Institute, was founded more than thirty years ago with an initial focus on the U.S.-Soviet relationship and arms control, but since then has evolved to examine the most critical foreign policy and national security issues confronting the country with a perspective informed by economic, social, and transnational considerations. Nowhere is this shift more evident than in our selection of cybersecurity as a topic for the Summer Workshop. Thirty years ago, when this group first met, we could not imagine the types of challenges we would face in the cyber domain. But it is thanks to the dynamism of the ASG that we can extend our modus operandi of open and frank dialogue to issues that are truly at the forefront of national security.

In the face of one of the most complex threats to national security, we believe that there are encouraging signs of American resilience and capability: the government has prepared a strategy and taken creative action, coordination across agencies seems to be increasingly managed, and the U.S. has the advantage of possessing highly advanced and competent professionals in both the public and private sector. But other states and non-state actors are also exploiting these challenges, assessing America's vulnerability and developing their own capabilities, norms, and plans. Storm clouds are looming on the horizon and, in a networked world, everyone must be vigilant to address this shared threat while preserving the benefits of an open Internet. We hope this book helps shed some light on the dangers ahead and the contributions we can each make to ensure a more secure cyberspace.

Preface

Nicholas Burns
Director, Aspen Strategy Group
Professor of the Practice of Diplomacy and International Politics
John F. Kennedy School of Government
Harvard University

The array of dangers from cyberspace—cybercrime, cyberterrorism and cyberespionage—have become as difficult and complex a set of national security challenges as the United States and other democratic countries have faced in the last generation. In just a few years, these cyber issues have burst onto the international scene as major dilemmas for policymakers in every major country of the world. The invention of the Internet, which facilitates unprecedented access to information, ease of communication, and unfettered opportunities for free expression, has become a defining and revolutionary feature of our age. For most of the last quarter century, governments and individuals focused largely on the many benefits of the information age and the extraordinary change it brought to our economy and way of life. Dangers from the cyber domain were not on the radar screens of most global leaders a decade ago, but now are front and center as a current and future preoccupation. Leaders are now scrambling to educate themselves and their publics on this complicated and multifaceted subject, to raise defenses, and to contemplate how to go on the offensive against cybercriminals.

The members of the Aspen Strategy Group met over five days in early August 2011 in Aspen, Colorado to examine the many dimensions of "Securing Cyberspace: A New Domain for National Security." We are a non-partisan group of former senior government officials, business leaders, academics, and journalists who gather several times per year to debate and discuss the great international challenges to American foreign policy. Founded thirty years ago by former National Security Advisor Brent Scowcroft and Harvard Professor Joseph Nye, our group now includes, among others, former Secretaries of State Madeleine Albright and Condoleezza Rice, former Secretary of Defense Bill Perry, former National Security Advisor Sandy Berger, award-winning journalists David Ignatius, David Sanger, and Nick Kristof, and business leaders Steve Friedman and Tom Pritzker. The Aspen Strategy Group's mission is to focus on difficult problems across party and ideological lines and to develop practical strategies for American success in the world.

Inspired by our co-chair, Joseph Nye, we decided that the Strategy Group should investigate the extraordinarily complicated and imposing threats emanating from the Cyber Domain. We were joined by many of the most knowledgeable experts in the field and several senior government officials—National Security Agency Director General Keith Alexander, Deputy Secretary of Defense Bill Lynn, State Department Cyber Issues Coordinator Christopher Painter, and Deputy Secretary of Homeland Security Jane Holl Lute—responsible for developing an American strategy on this critical issue.

We discovered in Aspen just how pervasive and forbidding the many cyber threats are to our government, the private sector, and American citizens.

Cybercrime is a problem for the majority of American businesses and accounts for billions of dollars in losses to our economy annually. We should assume that every sector of our economy has already been penetrated by criminals operating in the darkness and relative anonymity of cyberspace.

Cyberespionage is a significant new frontier in the long-running competition among states for influence and secrets that poses extraordinary challenges to preparing adequate defenses.

Cyberterrorism may be the most pernicious threat as it empowers groups of individuals with relative anonymity to plan lethal attacks against unsuspecting and often defenseless citizens as well as their governments.

You will find in this book's individual chapters compelling and thoughtful presentations on the cyber challenge that amount to a wake-up call for all Americans as well as our government. Among the most important issues are the following:

--First, led by the U.S., governments are rapidly mobilizing to get smarter on the dimensions of the cyber threat and to prepare to combat it. In the U.S., the Department of Defense is far ahead of most government agencies in the establishment of Cyber Command and in deciding to classify cyberspace as the fifth domain of warfare.

--Second, the rest of the U.S. government must act much more resolutely to get smarter about the problem and take action to combat it. Our government must act to protect our critical infrastructure—such as banks, the electric grid, and communications—from potential attack.

--Third, the private sector is relatively unaware and ill-prepared to protect the heart of our economy—small businesses. A much greater effort must be made by the American business community to recognize the threat and begin to counteract it.

--Fourth, individual Americans are subject to cyberattacks on their home computers and other electronics where their personal and financial information is in danger of cyber theft; in turn everyone has a role to play in cyber defense.

--Fifth, we are in the early stages of the campaign to secure cyberspace where the offensive threat of the cyberterrorists, criminals, and spies is likely much further advanced than our defenses. We need to work urgently to build adequate defenses and then launch a campaign between government and business to go on the attack against cybercriminal networks.

This book's chapters focus on these major threats and illustrate the complexity of the dangers from the cyber domain and what we can do about them. Our authors pose difficult questions for our national and business leaders, including:

--How should we convince China and Russia, major suspected cybercriminal nations, to cease and desist in sponsoring or allowing cyber threats to emanate directly from their territories?

--How can our own government provide more clear, resolute, and effective guidance and regulation to businesses and citizens? Some of our authors allege that the U.S. government is, in its passivity and inaction, essentially an accomplice to the crimes being committed against us.

--How can our major banks and businesses adopt a more open, transparent, and proactive stance to warn consumers of the threats to our financial system? Can we motivate business to share information on cyberattacks with each other and the government so they can learn from one another?

--Finally, how can we organize both here at home and across the world to build a global network to better secure the financial system, bank accounts, personal information, and government secrets from the international criminal networks preying on them in cyberspace? And, how can we do so without endangering the privacy and civil rights of our own citizens in an age of Internet freedom?

We think you will agree that this is an important issue for all Americans and for people from all law-abiding countries. If nothing else, we hope this book will ring the village bell to warn our country and others of the present and future dangers within the cyber domain.

Part  I

CYBERWARS & CYBERTERROR: UNDERSTAND-
ING CYBERSPACE AS A NEW BATTLEGROUND

The Third Annual Ernest May Memorial Lecture

Nuclear Lessons for Cybersecurity?

Joseph S. Nye, Jr.
University Distinguished Service Professor
John F. Kennedy School of Government
Harvard University

CHAPTER 1

U.S. Cybersecurity: The Current Threat and Future Challenges

Eric Rosenbach
Principal
Good Harbor Consulting

Robert Belk
Politico-Military Fellow
John F. Kennedy School of Government
Harvard University

CHAPTER 2

Resilience, Disruption, and a "Cyber Westphalia": Options for National Security in a Cybered Conflict World

Chris Demchak
Professor
U.S. Naval War College Strategic Research Department

"It is worth examining the uneven and halting history of nuclear learning to alert us to some of the pitfalls and opportunities ahead in the cyber domain. Ernest May once described U.S. defense policy and the development of nuclear strategy in the first half decade following World War II as "chaotic." He would likely apply the same term to the situation in cyberspace today."

—JOSEPH S. NYE, JR.

The Third Annual Ernest May Memorial Lecture

Nuclear Lessons for Cybersecurity?[1]

Joseph S. Nye, Jr.
University Distinguished Service Professor
John F. Kennedy School of Government
Harvard University

Editor's Note: Professor Joseph S. Nye, Jr. presented the annual Ernest May Memorial Lecture at the Aspen Strategy Group's August 2011 workshop in Aspen, Colorado. The Ernest May Memorial Lecture is named for Ernest May, an international relations historian and Harvard John F. Kennedy School of Government professor, who passed away in 2009. ASG developed the lecture series to honor Professor May's celebrated lectures.

Identifying "revolutions in military affairs" is somewhat arbitrary, but some inflection points in technological change are larger than others: for example, the gunpowder revolution in early modern Europe, the industrial revolution of the nineteenth century, the second industrial revolution of the early twentieth century, and the nuclear revolution in the middle of the last century.[2] In this century, we can add the information revolution that has produced today's extremely rapid growth of cyberspace. Earlier revolutions in information technology, such as Gutenberg's printing press, also had profound political effects, but the current revolution can be traced to Moore's law and the thousand-fold decrease in the costs of computing power that occurred in the last quarter of the twentieth century.

Technology is a double-edged sword. It eventually becomes available to adversaries who may have primitive capabilities, but those same adversaries are also less vulnerable to dependence on advanced technologies. Four decades ago, the Pentagon created the Internet, and today, by most accounts, the U.S. remains the leading country in both its military and societal use. At the same time, however, because of greater dependence on networked computers and communication, the

U.S. is more vulnerable to attack than many other countries, and the cyber domain has become a major source of insecurity that can be utilized by America's enemies.[3]

Political leaders and analysts are only beginning to come to terms with this transformative technology. Until now, the issue of cybersecurity has largely been the domain of computer experts and specialists. When the Internet was created forty years ago, this small community was like a virtual village of people who knew each other, and they designed an open system with little attention to security. The commercial Web is two decades old, but it has exploded from a mere 10 million users in the early 1990s to some two billion users today. This burgeoning interdependence has created great opportunities and great vulnerabilities which strategists do not fully comprehend. As General Michael Hayden, former director of the CIA says, "rarely has something been so important and so talked about with less clarity and less apparent understanding…. I have sat in *very* small group meetings in Washington…unable (along with my colleagues) to decide on a course of action because we lacked a clear picture of the long-term legal and policy implications of *any* decision we might make."[4] When compared to the nuclear revolution in military affairs, strategic studies of the cyber domain are chronologically equivalent to 1960, but conceptually more equivalent to 1950. Analysts are still not clear about the meaning of offense, defense, deterrence, escalation, norms, arms control, or how they fit together into a national strategy.

What is Cyberspace?

Cyber is a prefix indicating computer and electro-magnetic spectrum related activities. The cyber domain includes the Internet of networked computers, but also intranets, cellular technologies, fiber optic cables, and space-based communications. Cyberspace has a physical infrastructure layer that follows the economic laws of rival resources and the political laws of sovereign jurisdiction and control. In this aspect, the Internet is not a traditional "commons." Cyberspace also has a virtual or informational layer with increasing economic returns to scale, and political practices that make jurisdictional control difficult. Attacks from the informational realm, where costs are low, can be launched against the physical domain, where resources are scarce and expensive. Conversely, control of the physical layer can have both territorial and extraterritorial effects on the informational layer. Cyber power can produce preferred outcomes *within* cyberspace or in other domains *outside* cyberspace. By analogy, sea power refers to the use of resources in the oceans domain to win naval battles on the ocean, but it also includes the ability to use the oceans to influence battles, commerce, and opinions on land.

The cyber domain is a complex manmade environment. Unlike atoms, human adversaries are purposeful and intelligent. Mountains and oceans are hard to move, but portions of cyberspace can be turned on and off by throwing a switch. It is cheaper and quicker to move electrons across the globe than to move large ships long distances through the friction of salt water. The costs of developing multiple carrier task forces and submarine fleets create enormous barriers to entry and make it possible to speak of American naval dominance. In contrast, the barriers to entry in the cyber domain are so low that non-state actors and small states can play significant roles at low levels of cost.

In *The Future of Power* I describe diffusion of power away from governments as one of the great power shifts of this century.[5] Cyberspace is a perfect example of this broader trend. The largest powers are unlikely to be able to dominate this domain as much as they have others like sea, air or space. While they have greater resources, they also have greater vulnerabilities, and at this stage in the development of the technology offense dominates defense in cyberspace. The United States, Russia, Britain, France, and China have greater capacity than other state and non-state actors, but it makes little sense to speak of dominance in cyberspace. If anything, dependence on complex cyber systems for support of military and economic activities creates new vulnerabilities in large states that can be exploited by non-state actors.

The term "cyberattack" covers a wide variety of actions ranging from simple probes, to defacing websites, denial of service, espionage, and destruction.[6] Similarly, the term "cyberwar" is used very loosely for a wide range of behaviors. In this, it reflects dictionary definitions of war that range from armed conflict to any hostile contention (for example, "war between the sexes" or "war on poverty"). At the other extreme, some use a very narrow definition of cyberwar as a "bloodless war" among states that consists only of conflict in the virtual layer of cyberspace. But this avoids important issues of the interconnection of the physical and virtual layers of cyberspace discussed above. A more useful definition of cyberwar is hostile actions in cyberspace that have effects that amplify or are equivalent to major kinetic violence.

In the physical world, governments have a near monopoly on large-scale use of force, the defender has an intimate knowledge of the terrain, and attacks end because of attrition or exhaustion. Both resources and mobility are costly. In the virtual world, actors are diverse, sometimes anonymous, physical distance is immaterial, and offense is often cheap. Because the Internet was designed for ease of use rather than security, the offense currently has the advantage over the defense. This might not remain the case in the long term, as technology evolves—including efforts at "re-engineering" some systems for greater security—but it remains the case at this stage. The larger party has limited ability to disarm or destroy the enemy, occupy territory, or effectively use counter-force strategies. Cyberwar, although only incipient at this stage, is the most dramatic of the potential threats. Major states with elaborate technical and human resources could, in principle, create massive disruption as well as physical destruction through cyberattacks on military as well as civilian targets. Responses to cyberwar include a form of inter-state deterrence (though different from classical nuclear deterrence), offensive capabilities, and designs for network and infrastructure resilience if deterrence fails. At some point in the future, it may be possible to reinforce these steps with certain rudimentary norms, but the world is at an early stage in such a process.

If one treats hacktivism as mostly a disruptive nuisance at this stage, there are four major categories of cyber threats to national security, each with a different time horizon and with different (in principle) solutions: Cyberwar and economic espionage are largely associated with states, and cybercrime and cyberterrorism are mostly associated with non-state actors. For the United States, at the present time, the highest costs come from espionage and crime, but over the next decade or so, war and terrorism may become greater threats than they are today. Moreover, as alliances and tactics evolve among different actors, the categories may increasingly overlap. In the view of Admiral Mike McConnell: "Sooner or later, terror groups will achieve cyber-sophistication. It's like nuclear proliferation, only far easier."[7] We are only just beginning to see glimpses of cyberwar—for instance as an adjunct in some conventional attacks, in the denial of service attacks that accompanied the conventional war in Georgia in 2008, or the recent sabotage of Iranian centrifuges by the Stuxnet worm. Deputy Defense Secretary William Lynn has described the evolution of cyberattacks from exploitation to disruption of networks, to destruction of physical facilities. He argues that while states have the greatest capabilities, non-state actors are more likely to initiate a catastrophic attack.[8] A "cyber 9/11" may be more likely than the often mentioned "cyber Pearl Harbor."

Nuclear Lessons for Cybersecurity?

Can the nuclear revolution in military affairs seven decades ago teach us anything about the current cyber transformation? At first glance, the answer seems to be no. The differences between the technologies are just too great. The National Research Council cites differences in the threshold of action and attribution—Nuclear explosions are unambiguous, while cyber intrusions that plant logic bombs in the infrastructure may go unnoticed for long periods before being used, and even then can be difficult to trace.[9] Even more dramatic is the sheer destructiveness of nuclear technology. Unlike nuclear, cyber does not pose an existential threat. As Martin Libicki points out, destruction or disconnection of cyber systems could return us to the economy of the 1990s—a huge loss of GDP—but a major nuclear war could return us to the Stone Age.[10] In that and other dimensions, comparisons of cyber with biological and chemical weaponry might be more apt.

Moreover, cyber destruction can be disaggregated, and small doses of destruction can be administered over time. While there are many degrees of nuclear destruction, all are above a dramatic threshold or firebreak. In addition, while there is an overlap of civilian and military nuclear technology, nuclear originated in war, and the differences in its use are clearer than in cyber, where the Web has burgeoned in the civilian sector. For example, the "dot mil" domain name is only a small part of the Internet, and 90 percent of military telephone and internet communications travel over civilian networks. Finally, because of the commercial predominance and low costs, the barriers to entry to cyber are much lower for non-state actors. While nuclear terrorism is a serious concern, the barriers for non-state actors gaining access to nuclear materials remain steep, while renting a botnet to wreak destruction on the Internet is both easy and cheap.

It would be a mistake, however, to neglect the past so long as we remember that metaphors and analogies are always imperfect.[11] Ernest May liked to quote Mark Twain's aphorism that history never repeats itself: but sometimes it rhymes. There are some important nuclear-cyber strategic rhymes, such as the superiority of offense over defense; the potential use of weapons for both tactical and strategic purposes; the possibility of first and second use scenarios; the possibility of creating automated responses when time is short; the likelihood of unintended consequences and cascading effects when a technology is new and poorly understood; and the belief that new weapons are "equalizers" that allow smaller actors to compete directly but asymmetrically with a larger state.[12]

Even more important than these technical and political similarities is the learning experience as governments and private actors try to understand a transformative technology—and adopt strategies to cope with it. While government reports warning about computer and Internet vulnerability date back to 1991 and the Pentagon recently released a new cybersecurity strategy, few observers would argue that the country has developed an adequate national strategy for cybersecurity. It is worth examining the uneven and halting history of nuclear learning to alert us to some of the pitfalls and opportunities ahead in the cyber domain. Ernest May once described U.S. defense policy and the development of nuclear strategy in the first half decade following World War II as "chaotic."[13] He would likely apply the same term to the situation in cyberspace today.

General Lessons

1. Expect continuing technological change to complicate early efforts at strategy.

At first, both fissile materials and atomic bombs were assumed to be scarce, and it was considered wasteful to use atomic bombs against any but counter-value targets—that is, cities. Bernard Brodie and others concluded in the important 1946 book *The Absolute Weapon* that superiority in numbers would not guarantee strategic superiority; deterrence of war was the only rational military policy; and ensuring survival of the retaliatory arsenal was crucial.[14] These postulates of "finite" or "existential" deterrence persisted throughout the Cold War and serve as the basis for the nuclear strategies of countries such as France and China to this day. In the bipolar competition of the Cold War, however, the strategy of finite deterrence was challenged by the development of the hydrogen bomb in the early 1950s. Destructive power was no longer scarce, but now unlimited. While hydrogen bombs could lead to explosions counted in the tens of megatons, their real revolutionary effect was to permit miniaturization, which allowed multiple weapons to pack huge destructive power into the nose cones of another technological surprise, intercontinental missiles, which shortened response times to less than an hour. This burgeoning explosive power produced great concern about the vulnerability of limited arsenals, an enormous increase in the number of weapons, diminished prospects for active defenses, and the development of elaborate counter-force war-fighting strategies.

Both superpowers had to confront the "usability paradox." If the weapons could not be used, they could not deter. The U.S. and USSR were locked in a positive sum game that involved avoiding nuclear war, but simultaneously they were locked in a zero sum game of political competition. In the game of political chicken, perceptions of credibility became crucial. Some prospect of usability had to be introduced into doctrine, and for decades strategists wrestled with issues of counterforce targeting, exploring strategic defense technology, and the issues of perception that disparities in large numbers might create for extended deterrence. Elaborate war-fighting schemes and escalation ladders were invented by a nuclear priesthood of experts who specialized in arcane and abstract formulas. In 1976, Paul Nitze and the Committee on the Present Danger expressed alarm about American weakness when the U.S. possessed tens of thousands of weapons, and in 1979, even Henry Kissinger predicted that because of American nuclear weakness Soviet risk-taking "must exponentially increase."[15] In fact, the opposite proved to be the case. While politicians and strategists assailed the idea of mutual assured destruction as an immoral and dangerous strategy, MAD turned out to be a fact, not a policy. As McGeorge Bundy noted in his final work, when it came to the Cuban Missile Crisis, existential deterrence worked, and a few Soviet bombs created deterrence, despite an overwhelming American superiority in numbers.[16]

Looking at today's cyber domain, interdependence and vulnerability are twin facts that are likely to persist, but we should expect further technological change to complicate early strategies. ARPANET was created in 1969, and the domain name system and the first viruses date back to 1983, but as noted above, the mass use and commercial development of cyberspace date only from the invention of the Web in 1989 and widely available browsers in the mid-1990s.[17] As one expert put it, "as recently as the mid-1990s, the Internet was still essentially a research tool and the plaything of a few."[18] In other words, the massive vulnerabilities that have created the security problems we face today are less than two decades old and are likely to increase. While some experts talk about reducing vulnerability by re-engineering the Internet to make attribution of attack easier, this will take time. Even more important, it will not close all vectors of attack.

Early strategies focused on the network: improving code, computer hygiene, addressing issues of attribution, and maintaining air gaps for the most sensitive systems. These steps remain important components of a strategy, but they are far from sufficient. In some ways, the invention and explosion in the usage of the Web is analogous to the hydrogen revolution in the nuclear era. By leading society and the economy to a vast dependence on networked communications, it created enormous vulnerabilities that could be exploited not only through the Internet, but through supply chains, devices to bridge air gaps, human agents, and manipulation of social networks.[19] With the development of mobility, cloud computing and the importance of a limited number of large providers, the issues of vulnerability may change again. Given such technological volatility, a cybersecurity strategy will have to be multifaceted and capable of continual adaptation. It should increase the ratio of work that an attacker must do compared to that of a defender, and include redundancy and resilience to allow graceful degradation of complex systems so that inevitable failures are not catastrophic.[20] Strategists need to be alert to the fact that today's solutions may not suffice tomorrow.

2. Strategy for a new technology will lack adequate empirical content.

Since Nagasaki, no one has seen a nuclear weapon used in war. As Alain Enthoven, one of Robert McNamara's "whiz kids" of the early 1960s, retorted during a Pentagon argument about war plans, "General, I have fought just as many nuclear wars as you have."[21] With little empirical grounding, it was difficult to set limits or test strategic formulations. Elaborate constructs and prevailing political fashion led to expensive conclusions based on abstract formulas and relatively little evidence. As Fred Kaplan notes:

> The method of mathematical calculation, driven mainly from the theory of economics that they had all studied, gave the strategists of the new age a handle on the colossally destructive power of the weapons they found in their midst. But over the years the method became a catechism....The precise calculations and the cool, comfortable vocabulary were coming all too commonly to be grasped not merely as tools of desperation but as genuine reflections of the nature of nuclear war.[22]

In the absence of empirical evidence, these nuclear theologians were able to spend vast resources on their hypothetical scenarios.

Cyber has the advantage that with widespread attacks by hackers, criminals, and spies, there is more cumulative evidence of a variety of attack mechanisms and of the strengths and weaknesses of various responses to such attacks. It helps that cyber destruction can be disaggregated in a way that nuclear cannot. But at the same time, no one has yet seen a cyberwar in the strict sense of the word, as defined above. Denial of service attacks in Estonia and Georgia, and industrial sabotage such as Stuxnet in Iran give some inklings of the auxiliary use of cyberattacks, but they do not test the full set of actions and reactions in a cyberwar between states. The U.S. government has conducted a number of war games and simulations, and is developing a cyber test range, but the problems of unintended consequences and cascading effects have not been experienced. The problems of escalation as well as the implications for the important doctrines of discrimination and proportionality under the Law of Armed Conflict remain unknown.

3. New technologies raise new issues in civil-military relations.

Different parts of complex institutions like governments learn different lessons at different paces, and new technologies set off competition among bureaucracies. At the beginning of the nuclear era, political leaders developed institutions to maintain civilian control over the new technology, creating an Atomic Energy Agency separate from the military as a means of ensuring civilian control. Congress, meanwhile, established a Joint Atomic Energy Committee. But gaps still developed in the relationship between civilians and the military. Operational control of deployed nuclear weapons came under the Strategic Air Command (SAC), which had its own traditions, standard operating procedures, and a strong leader, Curtis LeMay. In 1957, LeMay told Robert Sprague, the deputy director of the civilian Gaither Committee that was investigating the vulnerability of American nuclear forces, that he was not too concerned because "if I see that the Russians are amassing their planes for an attack, I'm going to knock the shit out of them before they take off the ground." Sprague was thunderstruck and replied, "But General LeMay, that is not national policy," to which LeMay replied, "I don't care. It's my policy. That's what I'm going to do."[23] In 1960, when President Eisenhower ordered the development of a Single Integrated Operational Plan (SIOP-62), SAC produced a plan for a massive strike with 2,164 megatons that targeted "the Sino-Soviet Bloc."[24] The limited nuclear options that civilian strategists theorized about as part of a bargaining process would not have looked very limited from the point of view of the Soviet bargaining partner—not to mention China.

While Cyber Command is still new and has very different leadership from the old Strategic Air Command, cybersecurity does present some similar problems of relating civilian control to military operations. Time is even shorter. Rather than the 30 minutes of nuclear warning and possible launch under attack, today there would be 300 milliseconds between a computer detecting that it was about to be attacked by hostile malware and a preemptive response to disarm the attack. This requires not only advanced knowledge of malware being developed in potentially hostile systems, but also an automated response. What happens to the human factor in the decision loop? Obviously, there is no time to go up the chain of command, much less convene a deputies meeting at the White House. For active defense to be effective, authority will have to be delegated under carefully thought-out rules of engagement developed in advance. Moreover, there are important questions about when active defense shades into retaliation or offense. As the head of Cyber Command has testified, such legal authorities and rules still remain to be fully worked out. [25]

4. Civilian uses will complicate effective national security strategies.

Nuclear energy was first harnessed for military purposes, but it was quickly seen as having important civilian uses, as well. In the early days of the development of nuclear energy, it was claimed that electricity would become "too cheap to meter" and cars would be fueled for a year by an atomic pellet the size of a vitamin pill.[26] The engineers' optimism about their new technology was reinforced by a political desire to promote the civilian uses of nuclear energy. Fearful that anti-war and anti-nuclear movements would delegitimize nuclear weapons and thus reduce their deterrent value, the Eisenhower administration promoted an Atoms for Peace Program that offered to assist in the promotion of nuclear energy worldwide. Other countries joined in. The net effect was to create a powerful domestic and transnational lobby for promotion of nuclear energy that helped provide India with the materials needed for its nuclear explosion in 1974 and justified the French sale of a reprocessing plant to Pakistan and a German sale of enrichment technology to Brazil in the mid-1970s.

The Atomic Energy Commission and the Joint Atomic Energy Committee had been created to assure civilian control of nuclear technology, but over time both institutions became examples of regulatory capture by powerful commercial interests more interested in promotion than regulation and security. Late in the Ford administration, both institutions were disbanded. However, after the oil crisis of 1974 it became an article of faith that nuclear would be the energy of the future, that uranium would be scarce, and thus widespread use of plutonium and breeder reactors would be necessary. When the Carter administration, following the recommendations of the non-governmental Ford-Mitre Report,[27] tried to slow the development of this plutonium economy in 1977, it ran into a buzz saw of reaction not only overseas, but from the nuclear industry and its Congressional allies at home.

As discussed above, the civilian sector plays an even larger role in the cyber domain and this enormously complicates the problem of developing a national security strategy. The Internet has become a much more significant contributor to GDP than nuclear energy ever was. The private sector is more than a constraint on policy, it is at the heart of the activity that policy is designed to protect. Risk is inevitable, and redundancy and resilience after attack must be built into a strategy. Most of the Internet and its infrastructure belong to the private sector, and the government has only modest levers to use. Proposals to create a central agency in the executive branch and a Joint Committee on Cyber Security in Congress might be useful, but one should be alert to the dangers of regulatory capture and the development of a cyber "iron triangle" of executive branch, Congressional and industry partners.

From a security perspective, there is a misalignment of economic incentives in the cyber domain.[28] Firms have an incentive to provide for their own security up to a point, but competitive pricing of products limits that point. Moreover, firms have a financial incentive not to disclose intrusions that could undercut public confidence in their products and stock price. "The public (and very often the industry) understanding of this significant national security threat is largely minimal due to the very limited number of voluntary disclosures by victims of intrusion activity."[29] The result is a paucity of reliable data and an underinvestment in security from the national perspective. Moreover, laws designed to ensure competition restrict cooperation among private firms, and the difficulty of ascertaining liability in complex software limits the role of the insurance market. Public-private partnerships are limited by different perspectives and mistrust. As one participant at a recent cybersecurity conference concluded, something bad will have to happen before markets begin to re-price security.[30]

International Cooperation Lessons

1. Learning can lead to concurrence in beliefs without cooperation.

Governments act in accordance with their national interests, but they can change how they define their interests, both through adjusting their behavior to changes in the structure of a situation, as well as through transnational and international contacts and cooperation. In the nuclear domain, the initial learning led to concurrence of beliefs before it led to contacts and cooperation. The first effort at arms control, the Baruch Plan of 1946, was rejected out of hand by the Soviet Union as a ploy to preserve the American monopoly, and the early learning was unilateral on both sides.

As we have seen, much of what passed for nuclear knowledge in the early days was abstraction based on assumptions about rational actors, which made it difficult for new information to alter prior beliefs. Yet gradually both sides became increasingly aware of the unprecedented destructive power of nuclear weapons through weapons tests and modeling, particularly after the invention of the hydrogen bomb. As Winston Churchill put it in 1955, "the atomic bomb, with all its terrors, did not carry us outside the scope of human control," but with the hydrogen bomb "the entire foundation of human affairs was revolutionized."[31] In his memorable phrase, "safety became the sturdy child of terror."[32] On the other side of the Iron Curtain, Nikita Khrushchev said: "When I was appointed First Secretary of the Central Committee and learned all the facts about nuclear power I couldn't sleep for several days. Then I became convinced that we could never possibly use these weapons, and I was able to sleep again. But all the same we must be prepared."[33] These parallel lessons were learned independently. It was not until 1985 that Reagan and Gorbachev finally declared jointly that "a nuclear war cannot be won and must never be fought." That crucial nuclear taboo has lasted for nearly seven decades, and it was well ensconced before it was jointly pronounced.

A second area where common knowledge developed concurrently was in the command and control of weapons and the dangers of escalation, as the two governments accumulated experience of false alarms and accidents. A third area related to the spread of nuclear weapons: Both the U.S. and the Soviet Union gradually realized that sharing nuclear technology and expecting that exports could remain purely peaceful was implausible. A fourth area of common knowledge concerned the volatility of the arms race and the expenses and risks that it entailed. These views developed independently and in parallel, and it was more than two decades before they led to formal cooperation.

By its very nature, the interconnected cyber domain requires a degree of cooperation, and governments have become aware of this situation. Some analysts see cyberspace as analogous to the ungoverned Wild West, but unlike the early days of the nuclear domain, cyberspace has a number of areas of private and public governance. Certain technical standards related to Internet protocol are set (or not) by consensus among engineers involved in the non-governmental Internet Engineering Task Force (IETF), and the domain name system is managed by the Internet Corporation for Assigned Names and Numbers (ICANN). The UN and the International Telecommunication Union (ITU) have tried to promulgate some general norms, though with limited success. National governments control copyright and intellectual property laws, and try to manage problems of security, espionage, and crime within national policies. Though some cooperative frameworks exist, such as the European Convention on Cyber Crime, it remains weak, and states still focus on the zero sum rather than positive sum aspect of these games. At the same time, a degree of independent learning may be occurring on some of these issues. For example, Russia and China have refused to sign the Convention on Cyber Crime and have hidden behind plausible deniability as they have encouraged intrusions by "patriotic hackers." Their attitudes may change, however, if costs exceed benefits. For example, "Russian cyber-criminals no longer follow hands-off rules when it comes to motherland targets, and Russian authorities are beginning to drop the laissez-faire policy."[34] And China is independently experiencing increased costs from cybercrime. As in the nuclear domain, independent learning may pave the way for active cooperation later.

2. Learning is often lumpy and discontinuous.

Large groups and organizations often learn by crises and major events that serve as metaphors for organizing and dramatizing diverse sets of experiences. The Berlin crises and particularly the Cuban Missile Crisis of the early 1960s played such a role. Having come close to the precipice of war, both Kennedy and Khrushchev drew lessons about cooperation. It was shortly after the Cuban Missile Crisis that Kennedy gave his American University speech that laid the basis for the atmospheric test ban discussions.

Of course crises are not the only way to learn. The experience of playing iterated games of prisoner's dilemma in situations with a long shadow of the future may lead players to learn the value of cooperation in maximizing their payoffs over time.[35] Early steps in cooperation in the nuclear domain encouraged later steps, without requiring a change in the competitive nature of the overall relationship. These governmental steps were reinforced by informal "Track Two" dialogues such as the Pugwash conferences.

Thus far there have been no major crises in the cyber domain, though the denial of service attacks on Estonia and Georgia and the Stuxnet attack on Iran give hints of what might come. As mentioned earlier, some experts think that markets will not price security properly in the private sector until there is some form of visible crisis.[36] But other forms of learning can occur. For example, in the area of industrial espionage, China has had few incentives to restrict its behavior because the benefits far exceed the costs. Spying is as old as human history and does not violate any explicit provisions of international law. Nevertheless, at times governments have established rules of the road for limiting espionage and engaged in patterns of tit for tat retaliation to create an incentive for cooperation. While it is difficult to envisage enforceable treaties in which governments agree not to engage in espionage, it is plausible to imagine a process of iterations (tit for tat) which develops rules of the road that could limit damage in practical terms. To avoid "defection lock-in," which leads to unwanted escalation, it helps to engage in discussions that can develop common perceptions about red lines, if not fully agreed norms, as gradually developed in the nuclear domain after the Cuban Missile Crisis.[36] Discussion helps to provide a broader context (a "shadow of the future") for specific differences, and it is interesting to note that China and the U.S. have begun to discuss cyber issues in the context of their broad annual Strategic and Economic Dialogue, as well as in informal "Track Two" settings.

3. Learning occurs at different rates in different issues of a new domain.

While the U.S.-Soviet political and ideological competition limited their cooperation in some areas, awareness of nuclear destructiveness led them to avoid war with each other and to develop what Zbigniew Brzezinski called "a code of conduct of reciprocal behavior guiding the competition, lessening the danger that it could become lethal."[37] These basic rules of prudence included no direct fighting, no nuclear use, and communication during crisis. More specifically, it meant the division of Germany and respect for spheres of influence in Europe in the 1950s and early 1960s, and a compromise on Cuba. On the issue of command and control, concerns about crisis management and accidents led to the Hot Line, as well as the Accidents Measures and Incidents at Sea meetings of the early 1970s. Similarly, on the issue of non-proliferation the two sides discovered a common interest and began to cooperate in the mid-1960s, well before the bilateral arms control agreements about issues of arms race stability in the 1970s. Unlike the view that says nothing is settled in a deal until everything is settled, nuclear learning and agreements proceeded at different rates in different areas.

The cyber domain is likely to be analogous. As we have seen, there are already some agreements and institutions that relate to the basic functioning of the Internet, such as technical standards as well as names and addresses, and there is the beginning of a normative framework for cybercrime. But it is likely to take longer before there are agreements on contentious issues such as cyber intrusions for purposes like espionage and preparing the battlefield. Nevertheless, the inability to envisage an overall agreement need not prevent progress on sub-issues. Indeed, the best prospects for success may involve disaggregating the term "attacks" into specific actions that can be addressed separately.

4. Involve the military in international contacts.

As mentioned above, the military can be under civilian control but still have an independent operational culture of its own. By its nature and function, it is charged with entertaining worst case assumptions: It does not necessarily learn the same lessons at the same rate as its civilian counterparts. Early in the SALT talks, Soviet military leaders complained about the American habit of discussing sensitive military information in front of the civilian members of the Soviet delegation. The practice had the effect of broadening communication within the Soviet side. At the same time, Soviet military leaders had little understanding of American institutions or the role of Congress and how that would affect nuclear issues. Their involvement in arms talks helped to produce a more sophisticated generation of young leaders. As Foreign Minister Andrei Gromyko put it, "it's hard to discuss the subject with the military, but the more contact they have with the Americans, the easier it will be to turn our soldiers into something more than just martinets."[38]

In the cyber domain, the Chinese People's Liberation Army plays a major role in recruitment, training and operations. China today provides more opportunities for PLA generals to have international contacts than was true for Soviet officers during the Cold War, but those contacts are still limited. Moreover, while political control over the Chinese military is strong, operational control is weak, as shown by a number of recent incidents. Indeed, seven of the nine members of the Standing Military Commission wear uniforms, and there is no National Security Council or equivalent agency to coordinate operational details across the government. The lessons from the nuclear era would suggest the importance of involving PLA officers in discussions of cyber cooperation.

5. Deterrence is complex and involves more than just retaliation.

Early views of deterrence in the nuclear era were relatively simple and relied on massive retaliation to a nuclear attack. Retaliation remained at the core of deterrence throughout the Cold War, but as strategists confronted the usability dilemma and the problems of extended deterrence, their theories of deterrence became more complex. While a second strike capability and mutual assured destruction may have been enough to prevent attacks on the homeland, they were never credible for issues at the low end of the spectrum of interests. Somewhere between these extremes lay extended deterrence of attacks against allies and defense of vulnerable positions such as Berlin. Nuclear deterrence was supplemented by other measures, such as forward basing of conventional forces, declaratory policy, changes of alert levels, and force movements.

Many analysts argue that deterrence does not work in cyberspace because of the problem of attribution, but that is also too simple. Inter-state deterrence through entanglement and denial still exists, even when there is inadequate attribution. Even when the source of an attack can be successfully disguised under a "false flag," other governments may find themselves sufficiently entangled in symmetrically interdependent relationships that a major attack would be counterproductive: Witness the reluctance of the Chinese government to dump dollars to punish the U.S. after it sold arms to Taiwan in 2010.[39] Unlike the single strand of military interdependence that linked the U.S. and the Soviet Union during the Cold War, the United States, China, and other countries are entangled in multiple networks. China, for example, would itself lose from an attack that severely damaged the American economy, and vice versa.

In addition, an unknown attacker may be deterred by denial. If firewalls are strong, or the prospect of a self-enforcing response ("an electric fence") seems possible, attack becomes less attractive. Offensive capabilities for immediate response can create an active defense that can serve as a deterrent even when the identity of the attacker is not fully known. Futility can also help deter an unknown attacker. If the target is well protected, or redundancy and resilience allow quick recovery, the risk to benefit ratio in attack is diminished.[40] Moreover, attribution does not have to be perfect, and to the extent that false flags are imperfect, and rumors of the source of an attack are widely deemed credible (though not probative in a court of law), reputational damage to an attacker's soft power may contribute to deterrence. Finally, a reputation for offensive capability and a declaratory policy that keeps open the means of retaliation can help to reinforce deterrence. Of course, non-state actors are harder to deter, and improved defenses such as preemption and human intelligence become important in such cases. But among states, nuclear deterrence was more complex than it first looked, and that is doubly true of deterrence in the cyber domain.

6. Begin arms control with positive sum games related to third parties.

Although the U.S. and the Soviet Union developed some tacit rules of the road about prudent behavior early on, direct negotiation and agreements concerning arms race stability or force structure did not occur until the third decade of the nuclear era. Early efforts at comprehensive arms control like the Baruch Plan were non-starters. And even the eventual SALT agreements were of limited value in controlling numbers of weapons and involved elaborate verification procedures which themselves sometimes became issues of contention. The first formal agreement was the Limited Test Ban Treaty, where verification of atmospheric tests was easily detected and which could be considered largely an environmental treaty. The second major agreement was the Non-Proliferation Treaty of 1968, which was aimed at limiting the spread of nuclear weapons to third parties. Both these agreements involved positive sum games.

In the cyber domain, the global nature of the Internet requires international cooperation. Some people call for cyber arms control negotiations and formal treaties, but differences in cultural norms and the impossibility of verification makes such treaties difficult to negotiate or implement. Such efforts could actually reduce national security if asymmetrical implementation put legalistic cultures like the United States at a disadvantage compared to societies with a higher degree of government corruption. At the same time, it is not too early to explore international talks and cooperation. The most promising early areas for international cooperation are not bilateral conflicts, but problems posed by third parties such as criminals and terrorists.

For more than a decade, Russia has sought a treaty for broad international oversight of the Internet and "information security," banning deception or the embedding of malicious code or circuitry that could be activated in the event of war. But Americans have argued that arms control measures banning offense can damage defense against current attacks, and would be impossible to verify or enforce. Declaratory statements of "no first use" might have restraining effects on legalistic cultures like the United States while having less effect on states with closed societies. Moreover, the United States has resisted agreements that could legitimize authoritarian governments' censorship of the Internet. Cultural differences present a difficulty in reaching any broad agreements on regulating content on the Internet. The United States has called for the creation of "norms of behavior among states" that "encourage respect for the global networked commons," but as Jack Goldsmith has argued, "even if we could stop all cyberattacks from our soil, we wouldn't want to. On the private side, hacktivism can be a tool of liberation. On the public side, the best defense of critical computer systems is sometimes a good offense."[41] From the American point of view, Twitter and YouTube are matters of personal freedom; seen from Beijing or Tehran, they are instruments of attack. Trying to limit all intrusions would be impossible, but on the spectrum of attacks ranging from soft hacktivism to hard implanting of logic bombs in SCADA systems, one could start with cybercrime and cyberterrorism involving non-state third parties where major states would have an interest in limiting damage by agreeing to cooperate on forensics and controls. States might start with acceptance of responsibility for attacks that traverse their territory, and a duty to cooperate on forensics, information, and remedial measures.[42] In the future, it is possible that such cooperation could spread to state activities at the hard end of the spectrum, as it did in the nuclear domain.

Conclusion

Historical analogies are always dangerous if taken too literally, and the differences between nuclear and cyber technologies are great. The cyber domain is new and dynamic, but so was nuclear technology at its inception. It may help to put the problems of designing a strategy for cybersecurity into perspective, particularly the aspect of cooperation among states, if we realize how long and difficult it was to develop a nuclear strategy, much less international nuclear cooperation. Nuclear learning was slow, halting, and incomplete. The intensity of the ideological and political competition in the U.S.–Soviet relationship was much greater than that between the U.S. and Russia or the U.S. and China today. There were far fewer positive strands of interdependence in the relationship. Yet the intensity of the zero sum game did not prevent the development of rules of the road and cooperative agreements that helped to preserve the concurrent positive sum game.

That is the good news. The bad news is that cyber technology gives much more power to non-state actors than does nuclear technology, and the threats such actors pose are likely to increase. The transnational, multi-actor games of the cyber domain pose a new set of questions about the meaning of national security. Some of the most important security responses must be national and unilateral, focused on hygiene, redundancy, and resilience. It is likely, however, that major governments will gradually discover that cooperation against the insecurity created by non-state actors will require greater priority in attention. The world is a long distance from such a response at this stage in the development of cyber technology. But such responses did not occur until we approached the third decade of the nuclear era. With the World Wide Web only two decades old, may we be approaching an analogous point in the political trajectory of cybersecurity?

Joseph S. Nye, Jr. is University Distinguished Service Professor and former Dean of the John F. Kennedy School of Government, Harvard University. He is the Co-Chair of the Aspen Strategy Group. Dr. Nye has served as Assistant Secretary of Defense for International Security Affairs, Chair of the National Intelligence Council, and Deputy Under Secretary of State for Security Assistance, Science, and Technology. He published *The Powers to Lead* in March 2008. In 2004, he published *Soft Power: the Means to Success in World Politics*; *Understanding International Conflict* (5th edition); and *The Power Game: A Washington Novel*. His latest book, *The Future of Power* was published in February 2011. In a 2008 poll of international relations scholars, he was rated the sixth most influential over the past 20 years and the most influential on American foreign policy. He received a B.A. degree, *summa cum laude*, from Princeton University, did postgraduate work at Oxford University on a Rhodes scholarship, and earned a Ph.D. in Political Science from Harvard University.

[1] I am grateful to attendees of the August 2011 meeting of the Aspen Strategy Group for helpful comments on an earlier draft of this chapter.

[2] Oddly, Max Boot does not list the nuclear revolution in his discussion of revolutions in military affairs. See *War Made New: Technology, Warfare and the Course of History, 1500 to Today* (New York: Gotham Books, 2006).

[3] This point is emphasized by Richard A. Clarke and Robert Knake in *Cyberwar* (New York: Harper Collins, 2009).

[4] Gen. Michael V. Hayden, "The Future of Things Cyber," *Strategic Studies Quarterly* 5 (Spring 2011): 3.

[5] Joseph Nye, *The Future of Power* (New York: Public Affairs Press, 2011), chapter 5.

[6] For skeptical views that cyberwar is over-hyped, see Michael Hirsh, "Here There Be Dragons," *National Journal* (July 23, 2011): 32-7.

[7] McConnell quoted in Nathan Gardels, "Cyberwar," *New Perspectives Quarterly* 27 (April/Spring 2010): 16.

[8] Deputy Secretary of Defense William J. Lynn, III, "Remarks" (delivered at the 28th Annual International Workshop on Global Security, Paris, France, June 16, 2011). The text of the speech is available at: http://www.defense.gov/Speeches/Speech.aspx?SpeechID=1586.

[9] William Owens, Kenneth Dam and Herbert Lin, eds., *Technology, Policy, Law and Ethics Regarding U.S. Acquisition and Use of Cyberattack Capabilities* (Washington, D.C.: National Academies Press, 2009), 294.

[10] Martin C. Libicki, "Cyberwar as a Confidence Game," *Strategic Studies Quarterly* 5 (Spring 2011): 136. See also Libicki, *Cyberdeterrence and Cyberwar* (Santa Monica, CA: RAND, 2009), 136.

[11] Richard Neustadt and Ernest May, *Thinking in Time: The Uses of History for Decision-Makers* (New York: Free Press, 1986).

[12] Owens et al., 295-6.

[13] Ernest May, "Cold War and Defense," in Keith Neilson and Ronald G. Haycock, eds., *Cold War and Defense* (New York: Praeger, 1990), 54. I am indebted to Phillip Zelikow for bringing this to my attention.

[14] Fred Kaplan, *The Wizards of Armageddon* (New York: Simon and Schuster, 1983), 30.

[15] Kissinger quoted in Jervis, 102.

[16] McGeorge Bundy, *Danger and Survival: Choices About the Bomb in the First 50 Years* (New York: Vintage, 1990).

[17] Stuart Starr, "Toward a Preliminary Theory of Cyberpower," in Franklin Kramer, Stuart Starr, and Larry Wentz, eds., *Cyberpower and National Security* (Washington, D.C.: NDU Press, 2009), 82-6.

[18] Joel Brenner, *America the Vulnerable* (New York: Penguin Press, 2011), 15.

[19] On supply chain vulnerability, see Scott Charney and Eric Werner, *Cyber Supply Chain Risk Management: Toward a Global Vision of Transparency and Trust,* Microsoft, July 26, 2011.

[20] I am indebted to John Mallery of MIT CSAIL for his work on these points.

[21] Kaplan, 254.

[22] Ibid., 391.

[23] Ibid., 134.

[24] Ibid., 269.

[25] General Keith Alexander, quoted in "U.S Lacks People, Authorities to Face Cyber Attack," *Associated Press*, March 16, 2011.

[26] Brian Balogh, *Chain Reaction: Expert Debate and Public Participation in American Commercial Nuclear Power, 1945-1975* (Cambridge: Cambridge University Press, 1991), 31.

[27] The Nuclear Energy Policy Study Group, *Nuclear Power: Issues and Choices* (Ford-Mitre Report) (Cambridge, MA: Ballinger, 1977).

[28] See Brenner, *America the Vulnerable*.

[29] Dmitri Alperovitch, "Revealed: Operation Shady RAT," McAfee White Paper, 2011, 3.

[30] Jason Pontin, "Remarks" (delivered at the EastWest Institute Cyber Security Summit, London, UK, June 2, 2011).

[31] Churchill quoted in Michael Mandelbaum, *The Nuclear Revolution* (Cambridge: Cambridge University Press, 1981), 3.

[32] Ibid.

[33] Khrushchev quoted in Robert Jervis, *The Meaning of the Nuclear Revolution* (Ithaca: Cornell University Press, 1989), 20.

[34] Joseph Menn, "Moscow gets tough on cybercrime," *Financial Times*, March 22, 2010.

[35] See Robert Axelrod, *The Evolution of Cooperation* (New York: Basic Books, 1984).

[36] For a description of the gradual evolution of such learning in the nuclear area, see Joseph Nye, "Nuclear Learning and U.S.-Soviet Security Regimes," *International Organization* 41, 3 (Summer 1987). See also Graham Allison, "Primitive Rules of Prudence: Foundations of Peaceful Competition" in Allison and William Ury, eds., *Windows of Opportunity: From Cold War to Peaceful Competition in U.S.-Soviet Relations* (Cambridge, MA: Ballinger, 1989).

[37] Zbigniew Brzezinski, *Game Plan* (Boston: Atheneum, 1986), 244.

[38] Arkady Shevchenko, *Breaking With Moscow* (New York: Ballantine, 1985), 270-1. See also Raymond Garthoff, "Negotiating SALT," *The Wilson Quarterly* (Autumn 1977): 79.

[39] For details, see Chapter 3 in Nye, *The Future of Power.*

[40] I am indebted to the unpublished writings of Jeff Cooper on these points.

[41] Jack Goldsmith, "Can We Stop the Global Cyber Arms Race," *Washington Post*, February 1, 2010.

[42] See for example, Eneken Tikk, "Ten Rules for Cyber Security," *Survival* 53, 3 (June-July 2011): 119-132.

"The explosive growth of the Internet over the past decade created tremendous opportunities for the United States to improve its strategic position in the world: The Internet has enhanced nearly all aspects of American power."

—ERIC ROSENBACH & ROBERT BELK

U.S. Cybersecurity:
The Current Threat and Future Challenges

Eric Rosenbach
Principal
Good Harbor Consulting

Robert Belk
Politico-Military Fellow
John F. Kennedy School of Government
Harvard University

Much has been written about threats in cyberspace, but there has been little clarity in providing a sensible means of analyzing these threats. Moreover, though perceptions of these threats range from relatively naïve denial to overstated panic, there has been little illuminating analysis on the challenges that policymakers face in addressing the most likely concerns in cyberspace. This chapter attempts to rectify these shortcomings in two parts. The first section frames the fundamentals of the cyber threat, providing a sensible paradigm for understanding the types of threats. The second section builds upon this framework by outlining the four biggest future challenges policymakers face in attempting to improve U.S. cybersecurity.

PART I: A Cyber Threat Framework

Quantifying types of cyberattacks is particularly difficult, because the nature of cyber threats eschews clear lines of definitional demarcation. Threat types overlap, and attacks that cause equal effect are often not equivalent to each other in context. Before attempting to differentiate between different types of cyberattacks, therefore, it is prudent to first establish a working paradigm in order to clarify the context in which they take place.

Attempts to categorize cyber threats often depend on the perspective from which one views cyberspace. The framework shown in Figure 1 strives to compensate for these biases by providing a generic paradigm based on a function of motive and effect.

Motive for cyberattacks range from ideological to financial. Ideological motives are more general and can include philosophical tenets, personal predilection, and the quest for notoriety. Conversely, financially-motivated attacks are specifically rooted in cupidity, though value can broadly include simple immediate monetary reward as well as long-term benefit from intellectual property.

In contrast to motive, effect describes the direct physical effect of the cyberattack. Non-kinetic attacks would incorporate attacks that do not result in disruption or destruction of physical assets or active services. Kinetic attacks, however, represent attacks in which cyber technology is utilized to effect actual physical damage. In this way, effect also symbolizes the bridge between the virtual and physical worlds. Purely non-kinetic attacks target virtual domains, while purely kinetic attacks target physical ones.

Figure 1 illustrates the motive-effect function and the four major types of cyberattack. It is important to note that not only are these categories imprecise and the boundaries between them porous, but this matrix also implies that additional types of cyberattacks that are not incorporated in this framework exist. This version prioritizes simplicity over completeness and focuses on the four major categories of cyberattacks that are detailed in the following section.

Categories of Cyberattacks

The first major category of cyberattacks can be called cyber activism, or "hacktivism." Hacktivism is the use of cyber-centered tactics to further an ideological goal, make a political statement, or simply seek notoriety. As demonstrated in Figure 1, hacktivism is ideologically motivated with either non-kinetic or mildly kinetic effect.

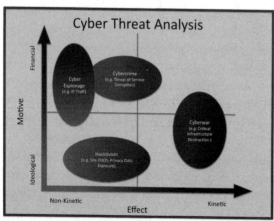

Figure 1: Cyber threat Analysis

Within these confines, hacktivists employ various tactics to achieve their objectives.

Compromising sensitive information involves the breaching of an organization's networks in order to steal, and subsequently distribute, sensitive personal information residing on the network. For example, in June 2011, to protest Arizona's strict anti-immigration laws, the hacktivist group LulzSec attacked the Arizona Department of Public Safety networks, and posted hundreds of sensitive documents online.

A slightly more "kinetic" tactic is the use of distributed denial of service (DDoS) attacks. DDoS attacks consist of flooding an organization's website with information requests from thousands of separate computers in order to overload the servers and effectively prevent online operations. For example, in December 2010, the hacktivist group Anonymous orchestrated a DDoS attack against MasterCard, Visa, and PayPal websites to protest these companies' lack of support for WikiLeaks.

Equally disruptive, but perhaps more akin to vandalism, website defacement entails compromising an organization's official website and inserting unauthorized content intended to shame the organization. For example, in July 2011, LulzSec infiltrated the public website of News Corp's *The Sun* tabloid and posted fake reports of CEO Rupert Murdoch's death to protest *The Sun's* hacking of crime victims' cell phone messages.

Regardless of the tactic employed, modern, public hacktivists claim that their operations are a form of non-violent protest that extend into, and utilize the unique mechanisms of, cyberspace. These ideologically aligned groups contrast to nationalistic hacktivists that support a country's ideals or values. The point where the latter groups begin to operate at the behest of government signals the limits of the definition of hacktivism. For example, a Russian crime network's DDoS attack against Georgian government websites in 2008 could be considered hacktivism, as the Russian government claimed. Yet, if executed at the request of the Russian government in coordination with a military attack on Georgia, this simple DDoS is now a tactical operation using cyberwarfare to augment traditional hard power (as will be detailed later).

The second major category of cyberattacks is cybercrime, which is unauthorized computer or network penetration with the intent of extortion or theft. As demonstrated in Figure 1, cybercrime is financially motivated with generally non-kinetic to mildly kinetic effects. Cybercrime is a particularly broad category of cyberattack, because there are many mechanisms with which an individual or group could perpetuate a crime (whether in cyberspace or not).

As one of the most pervasive forms of cybercrime, sensitive data theft is the use of cyber tactics to gain unauthorized access to sensitive data linked to individuals' financial information. For example, in May 2011 hackers gained access to Citigroup networks through a simplistic vulnerability in the company's banking website and stole the credit card information of over 360,000 clients. While sensitive data theft targets large reservoirs of personal information generally held by large corporations, individual financial theft targets information that is less localized. Individual financial thieves in cyberspace use phishing, spear-phishing, malware, and other cyber tactics to convince individuals to provide them with access to private online banking information. One example includes fraudulent spam emails from financial institutions that entice victims into providing the usernames and passwords to their accounts.

Though perhaps less frequently employed, cyber extortion is the threat to disrupt service, expose sensitive information, or damage or destroy critical infrastructure unless the perpetrator receives financial remuneration. For example, in June 2010, a German man rented a Botnet for $65 and threatened to disable gambling sites before the World Cup unless he was paid €2,500.

Cybercrime is characterized not just by a high risk/reward ratio, but also by the virtually instantaneous realization of its rewards. It is an extremely lucrative endeavor, spurred by the general anonymity of the attacks, the extensive victim pool, and in many cases the complicity of host nations in major cybercrime networks. The exact economic cost of cybercrime is difficult to quantify, because institutions have a strong disincentive to disclose breaches of their networks and share the repercussions of criminal exploitations. In July 2011, however, the National Security Council issued a report estimating the economic impact of organized cybercrime on American individuals and corporations at $1 billion per year.[1]

In contrast to cybercrime, cyberespionage is unauthorized computer penetration by state or non-state actors to obtain valuable information. As demonstrated in Figure 1, cyberespionage is ideologically and financially motivated with generally non-kinetic effects. Generally, there are two forms of cyberespionage based on the target of the espionage. Classified information-based cyberespionage involves breaching an organization's networks with the intent of obtaining classified documents and communiqués. This classified information includes state and foreign policy analyses as well as defense-related data. For example, in April 2009, attackers allegedly from China gained unauthorized access to extensive information pertaining to the U.S. Department of Defense's Joint Strike Fighter Program.

While classified information-based cyberespionage generally involves national documents, intellectual property-based cyberespionage targets corporations; breaching their networks with the intent of obtaining sensitive intellectual property. For example, the December 2009 Aurora attacks targeted source code from major technology companies including Google and Adobe. The 2009-2010 Night Dragon attacks targeted intellectual property from major energy, oil, and petrochemical companies. Both attacks allegedly originated in China.

Cyberespionage is an extremely complex form of cyberattack for three major reasons. First, unlike hacktivism and even organized cybercrime, cyberespionage is perpetrated by state-based, advanced, and persistent threats. These attackers are well funded, patient, and sophisticated. Second, the cost of cyberespionage is extremely high, though, like cybercrime, poorly quantified. Estimates run the gamut from hundreds of millions of dollars to over a trillion dollars annually. Third, being state-sponsored, cyberespionage is not simply a law-enforcement issue, but a diplomatic one. In some countries, cyberespionage against Western targets is not simply permissible, but laudatory.

The superficial distinction between cyberespionage and some forms of hacktivism is motive and sponsorship. While WikiLeaks has posted vast amounts of classified DoD data and the Anti Security (antisec) movement has supported unfettered information transparency, neither is state-sponsored. Moreover, neither is driven by financial motives or a desire to increase the relative power of their host state.

Lastly, cyberwar is a state's use of networks and digital techniques to disrupt, damage, or destroy a rival state's critical systems. As demonstrated in Figure 1, cyberwar is predominantly ideologically motivated, with mostly kinetic to purely kinetic effects. U.S. policy has attempted to define cyberwar, actions it believes constitute acts of cyberwar, and responses to such acts. Nevertheless, there has been little illumination of the impact of various cyberwar initiatives, and this lack of clarity obscures the natural question of proportionality of response.

Cyberwarfare, therefore, can be differentiated by the manner in which states leverage cyberweapons and their resultant effect. At the most basic level, battlefield preparation is a state's efforts to assess and position itself within a foreign country's networks, include probing vulnerabilities, mapping network infrastructure, and inserting logic bombs. For example, in 2009 the United States learned that hackers from Russia and China had exploited the U.S. electricity network and that certain portions were susceptible to remote disruption.

Combining knowledge gained through battlefield preparation, cyberwar can be a means of hard power support. This form of cyberwar is a state's use of cyber technology to disrupt digital, electronic, supporting, and network systems as an additional attack vector for traditional kinetic operations. For example, Israel allegedly used cyberattacks on Syrian air defenses in 2007 during the airstrike on a nuclear facility in Deir ez-Zor. Additionally, as mentioned above, DDoS and vandalizing attacks on Georgian government websites during the Russia-Georgia conflict in 2008 could be considered a form of hard power support.

Kinetic cyberwar is a state's use of cyber technology to destroy a foreign nation's militarized target or critical infrastructure systems. Such critical infrastructure may include financial institutions, transportation networks, and energy transmission systems. Though there is some debate about whether certain actions could be considered pure kinetic cyberwar (for example, the use of Stuxnet in Iran's Natanz nuclear centrifuges), there are still no confirmed cases of deliberate kinetic cyberwar. It is alleged, however, that in the early 1980s the CIA inserted code into Soviet natural gas systems that created unsafe operating parameters and resulted in a three-kiloton explosion in Siberia.

Though actual cyberwar may be theoretical, cyberwarfare is garnering increased attention as states begin to analyze the implications of operations in the rapidly developing battlespace. While Britain has enacted well-publicized spending cuts in defense, it has dramatically increased funding for cybersecurity. U.S. Defense Secretary Leon Panetta postulated at his confirmation hearing that the next Pearl Harbor-type attack could be through cyberwarfare.

The potential severity of the threat is underscored by the difficulty of formulating a cohesive strategy to combat it. Current complexities include attribution, deterrence, and proportionality of response. In addition, because most of the Western nation networks are privately owned, there are legal and regulatory issues that increase the complexity of preparing for cyberwar.

Part II: Four Big Challenges For Cybersecurity Policymakers

With the framework above serving as context, it is now possible to assess the challenges the United States faces with regards to cybersecurity. These challenges have developed coincidentally with the expansion of networked communications and globalization. The explosive growth of the Internet over the past decade created tremendous opportunities for the United States to improve its strategic position in the world: The Internet has enhanced nearly all aspects of American power. On the economic front, a recent report by the McKinsey Global Institute predicts that the Internet will continue to be the primary driver of American economic growth for the next decade. The emergence of sophisticated social networking and fundraising tools has facilitated the growth of dynamic organizations dedicated to advancing the public good. From a military perspective, America's mastery of Internet-based communications and weapons provides distinct advantages to our forces in modern warfare.

In short, protecting key assets in cyberspace is a vital national interest; the Internet may well represent America's modern-day equivalent of the Clausewitzian "center of gravity." Both state and non-state actors realize that the most significant blow to the United States would be an attack that cripples the networks running the nation. Fortunately, a blow of that magnitude is unlikely to occur in the next decade. Unfortunately, our adversaries and rivals recognize that the ubiquitous and open nature of cyberspace allows them to attack the United States on a small, yet significant, scale each day.

When it comes to cybersecurity, defense is much more difficult than offense. The challenge of improving the United States' cyber defenses is so complex and multifaceted that leading national security minds naturally gravitate to more attractive issues related to offense, such as retaliatory and first-use policies. Moving forward, it would be helpful to think deeply about creative ways in which the United States can design policies that improve our cyber defenses. While "active defense" is a critical component of cybersecurity, even the best offense does not compensate for an even decent defense.

As detailed below, many of the policymaking challenges in cyberspace do not relate directly to traditional areas of national security policy. Important challenges in technology and economic policy, for example, will play a key role in shaping cybersecurity policy. Seen from this perspective, there are four main challenges that make policymaking in this area exceptionally difficult.

Challenge One: American policymakers must find technological solutions to close the vectors through which adversaries attack our interests in cyberspace.

The original rules of the technology that govern the Internet, known as Internet Protocol, were designed without security in mind. These standards succeeded in facilitating widespread growth and adoption of the Internet; however, several aspects of this open architecture complicate national security policymaking. Likewise, the American software industry has developed all of the "special sauce" that allows the Internet to so effectively meet our needs, yet malicious software (known as malware) plagues our systems. There are four main characteristics of cyberspace that help explain this phenomenon and that deserve special attention when it comes to cybersecurity policy.

The first is, quite simply, speed. In cyberspace, the speed of attacks is measured in milliseconds, not minutes. Systems and networks can come under attack at any time, from nearly any network, with little or no indication or warning. This complicates several aspects of security policy because past strategic thought rested on the assumption that a state had the ability to detect an impending attack. For example, U.S. nuclear response doctrine during the Cold War rested on the assumption that a missile launch from the Soviet Union could be detected prior to detonation on American soil. Under international public law, states are entitled to defend themselves from the use of force when an armed attack is imminent. In many ways, however, the notion of imminence is anachronistic when it comes to cyberwar.

Nuclear response doctrine also rested on the certainty of attribution. Attribution in cyberspace, however, is currently complicated at best, and perhaps even impossible. The Internet's open architecture allows many actors and systems to remain anonymous, rendering it difficult to definitively assign attribution to perpetrators of malicious acts. Additionally, many attackers in cyberspace use proxy servers to mask their physical location. For example, an attacker could be physically located in Moldova, establish a proxy server in China, use that proxy server to control a network of compromised computers (known as a botnet) in Germany, and direct the botnet to attack banks or critical infrastructure in the United States. Thus, determining an appropriate "mailing address" for a potential American response to an attack is complex and slow, at best.

Compounding the difficulty of attribution is the problem of software resilience. At a fundamental level, the basis of most cyberattacks is exploitable vulnerabilities due to errors in software code. A commonly accepted estimate among the technology community is that for every one thousand lines of code, there will be roughly one error. As computer technology has advanced over the past decade to perform increasingly sophisticated tasks, so too has the complexity of the software driving these systems. This results in software with massive amounts of code, and, consequently, many exploitable vulnerabilities. Consider the case of the operating system Microsoft Windows: Windows 95, first released in August 1995, had roughly 10 million lines of code; Windows XP, released six years later in 2001 had 40 million; Windows Vista, released in late 2006 and early 2007, contained 50 million. In just over ten years the amount of code behind the operating system increased by a factor of five.[2] The associated coding holes allow adversaries to design specialized "zero-day" exploits, software-based attacks that are nearly impossible to defend against because the cybersecurity community has never seen them before.

Lastly, there is the pervasive interconnectedness of modern society, exemplified by the smart grid. The smart grid is a broad term referring to many initiatives that use the Internet and systems control technology to improve the reliability, security, and efficiency of the electric power grid. The smart grid has the potential to greatly improve the performance of electrical grids and reduce energy consumption; however, the widespread connectivity of these systems introduces new cybersecurity vulnerabilities to electrical grids. In 2009, for example, the *Wall Street Journal* reported that Chinese cyber spies had penetrated the U.S. electrical grid and left behind software programs that could be used to disrupt the power grid. Cyberattacks on the electrical grid could not only black out large parts of the country, but also trigger the physical destruction of critical electrical production and distribution systems through the manipulation of control systems. A classified exercise conducted by the Department of Homeland Security in 2007 demonstrated that previously theoretical cyberattacks against power generation equipment were, in fact, achievable.

Challenge Two: The United States lacks an economic policy framework that will bolster our competitiveness by improving cybersecurity.

Most policy analysis in Washington focuses myopically on cyberwar and potentially catastrophic attacks against critical infrastructure. As noted above, however, cyberwar is by far the least likely form of cyberattack. The most significant risks to the United States in cyberspace are daily cyberattacks that undermine our national economic competitiveness. Cybercrime, most often perpetrated by organized syndicates, and cyberespionage, carried out on a massive scale by the Chinese, bleed the United States of tens of billions of dollars each year.[3] The White House recognizes this fact: President Obama recently argued that "America's economic prosperity in the twenty-first century will depend on cybersecurity."

Too few policymakers recognize that the private sector, not the government, is the essential player in cybersecurity. American companies design and operate nearly all of the information technology systems vital to the Internet and the U.S. economy. Most private sector executives genuinely understand the importance of protecting America's intellectual property, financial systems, and critical infrastructure. Moreover, a surprising number of executives proactively seek guidance from the National Security Agency on how they can contribute to national security, even at the expense of their bottom line. At the same time, however, few businesses want additional regulation or legal requirements related to cybersecurity. The private sector understandably fears government intervention in the midst of a shaky economy. In the services sector, businesses believe that increased regulatory measures will disproportionately place the costs of improving the nation's cybersecurity upon their shoulders. In the technology sector, firms believe that additional regulation will slow the development of new products.

The Obama administration believes strict regulatory action is not the best strategy for correcting cybersecurity woes. Larry Summers, the former Treasury secretary and White House economic advisor, has argued that our technology sector is our last great export industry. He believes that hampering it with cybersecurity requirements would place the nation at a comparative disadvantage to foreign developers and manufacturers. And clearly the prospect of a global agreement that would level the playing field by mandating certain types of security measures in technology products is distant, at best.

In many ways, however, this is a classic case of market failure that deserves government intervention. In many different sectors of the global economy, organizations and corporations who practice poor cybersecurity shift the costs of cyberattacks to others. Although several high-profile and expensive attacks in the last several months may crystallize the potential financial risks of poor cybersecurity, many manufacturers have little incentive to invest additional resources in improving the security of their products. For example, software makers sell products with porous code because the costs of insecure software rarely, if ever, come back to them. On the other hand, the financial services industry invests tremendous resources in cybersecurity because they are legally and financially responsible for most cybercrime perpetrated against their customers.

Clearly, policymakers must think carefully about creating incentives or regulation to stimulate security investments. In the current fiscally constrained economic environment, the more effective form of government intervention may be to develop certifications based on clearly defined standards, sharing research conducted by the intelligence community on cybersecurity vulnerabilities with technology makers and operators.

Challenge Three: The United States must clarify several ambiguous areas of the law that constrain our ability to improve the nation's cyber defenses.

The current legal framework that governs cybersecurity is a complex and highly nebulous patchwork of international, national, and state laws that exacerbate the challenge of securing cyberspace in two important ways. First, private sector operators of critical telecommunications infrastructure fear that many of their best options for mitigating attacks, such as more assertive network monitoring or information sharing, could result in costly civil litigation. Thus, Internet service providers (ISPs) make a conscious cost-benefit decision to allow certain types of malicious behavior to occur on the networks. Second, constantly evolving interpretations of the law of war cloud policymakers' understanding of conflict in cyberspace and minimize our ability to deter potential adversaries. Thus, efforts to develop a more effective legal framework should focus on electronic surveillance and the laws of cyberwar.

One of the most difficult legal challenges for cybersecurity will be balancing constitutional concerns about privacy with a desire to conduct proactive monitoring of networks. Much of the debate on this issue focuses on the question of whether an ISP, possibly in conjunction with the government, can monitor users on their networks to identify the types of cyberattacks that enable cybercrime, cyberespionage, and cyberwar. Because they control users' access to their networks, ISPs have the ability to scan devices and either notify a user that he is infected or block his access to the Internet. To date, only one national ISP, Comcast, has elected to proactively monitor networks for viruses and a particularly malicious type of attack based on a string of infected computers.

Most Americans instinctively object to proposals for electronic monitoring of Internet activity by either the government or corporations, even if under the auspices of improving their security. As previously mentioned, the original design of the Internet often makes it difficult to determine an individual's identity. Thus, as the Internet evolved, the expectation of anonymity quickly became an integral aspect of Internet culture, even if many forms of Internet activity have a lower level of protection under the law than some older forms of telecommunications.

Networks controlled by the federal government are already subject to monitoring under a program run by the Department of Homeland Security known as "Einstein." Many experts assert that unless the government expands this program to monitor the public Internet for potential threats, the nation will be essentially blind to potentially catastrophic cyberattacks. Recent media reports have emerged that the NSA has initiated a program called "Perfect Citizen" to monitor the networks connected to critical infrastructure, such as nuclear plants and the electric grid. Government surveillance of critical infrastructure connections to the Internet may already be a reality, then; but more widespread monitoring will certainly raise additional legal and policy challenges.

The difficulty associated with attributing cyberattacks and definitional confusion about the use of force complicate application of the law of war to cyberspace. As discussed above, positive attribution of responsibility for cyberattacks is difficult. Without certain knowledge of the location and identity of an attacker, the law of war does not permit launch of a retaliatory operation. More importantly, the international community and the United States are still grappling with two important questions related to the use of force in cyberspace.

First, the U.S. has not clearly articulated standards about the types of hostile enemy actions that would constitute an armed attack in cyberspace. While the White House and Pentagon have repeatedly asserted that a cyberattack against the U.S. could trigger a kinetic response, they have demurred on clearly defining this threshold. Use of force in conventional warfare is readily perceivable: The bombing of a power plant or military installation clearly crosses American red lines. Cyberattacks intended to sabotage the control systems of American power plants or communications systems, however, may also reach the threshold of armed attack, since the end effect and intent of the attack mirror those of an attack in the physical world. The United States should seek to explicitly highlight red lines that adversaries should not cross in cyberspace in order to bolster the impact and effectiveness of strategic deterrence.

Second, U.S. leaders must continue to clarify the rules governing the deployment of American offensive cyberweapons. In May 2011, the Pentagon developed a classified list of cyberweapons that could be used to penetrate and disrupt the networks of an adversary and integrated this list into the formal structure of approved weapons.[5] This is a major development in the military's cybersecurity doctrine. Henceforth, military personnel will hopefully have clear rules of engagement governing the use of cyberweapons, just like any other weapons system. In the words of one military official, "whether it's a tank, an M-16 or a computer virus, it's going to follow the same rules that help us understand how to employ it, when you can use it, when you can't, what you can and can't use."[6] *The Washington Post* also revealed that penetrating a foreign network to plant a logic bomb for activation at a later date would require presidential approval, but penetrating such systems for espionage purposes would not.

Challenge Four: The United States needs new, creative policies that will mitigate Chinese efforts to exploit cyberspace to advance their national interests.

No actor challenges American interests in cyberspace more strongly or directly than China, which has deftly integrated cyberspace into nearly all aspects of its national strategy. China has sought to bolster its military power by establishing a cyberwar unit, and in an effort to erode American dominance in the information technology sector, the Chinese have pursued an effective industrial policy by supporting telecommunications firms like Huawei and ZTE. But the true core of Chinese strategy in cyberspace focuses on acquiring valuable information.

From Google to major energy firms, from the Pentagon to Congress, Chinese spies continue to steal the intellectual property, military plans, and secret information that keep America strong. Alone, no one attack represents a major threat to our national security; together, however, they imperil our long-term strategic position in the world. Although the likelihood is very low that this type of Chinese aggression in cyberspace will lead to a full-scale cyberwar or kinetic conflict, American policymakers and strategists must think deeply about the implications of competition in this new domain.

Chinese military strategists like to remind Americans that, "Those who live in a glass house shouldn't throw stones."[7] The message is clear: The U.S. is extremely vulnerable to cyberattack because it is open, democratic, and highly networked. This aspect of Chinese cyberspace strategy emerged in conjunction with the United States' obliteration of Iraq in the Gulf War. After quickly recognizing that it would take several decades to build a military that could match the conventional strength of the U.S., the Chinese military turned to Sun Tzu's writings on asymmetric warfare. As former Director of National Intelligence Mike McConnell noted, "the Chinese concluded from the Desert Storm experience that their counter approach had to be to challenge America's control of the battlespace by building capabilities to knock out our satellites and invade our cyber networks."[8]

China's interest in leveraging cyberspace extends beyond the military. In its 2006 National Medium- and Long-Term plan for the Development of Science and Technology, the Chinese government articulated a clear goal to reduce domestic dependence on advanced technologies developed overseas, particularly in the U.S. and Japan. Bolstering domestic research and development spending on technology related to cyberspace is also a key priority. As Adam Segal of the Council on Foreign Relations has noted, the Chinese have pursued the goals laid out in the report in three ways: through industrial policy, innovation strategy, and cyberespionage.[9]

The Chinese have focused their espionage efforts on stealing the core of America's economic strength and competitiveness: intellectual property and corporate secrets. Chinese cyberespionage has hit one industry after another. In the most well-known instance, a sophisticated Chinese operation in 2009 stole the source code from leading U.S. information technology companies such as Google, Cisco, and Adobe. In another audacious operation, Chinese cyber spies stole the corporate secrets—including the locations of several newly discovered oil fields—of several major corporations in the energy sector. In sum, the Chinese economy is the direct beneficiary of the theft of billions of dollars-worth of research and development that had been paid for by stockholders and taxpayers in the United States.

China's cyberespionage operations also extend to the highest levels of American political leadership. In 2007, for example, the Pentagon revealed with "very high levels of confidence" that cyber spies from the People's Liberation Army had hacked into the computers in the office of Defense Secretary Robert Gates.[10] In 2008, experts confirmed that that the Chinese had penetrated and spied on the computers used by members of the House Foreign Affairs Committee, most likely because of their vocal criticism of China's human rights record. When Congressman Frank Wolf, a leading member of the committee, tried to mobilize the U.S. government to take action he quickly learned that "no one wants to talk about this issue."[11]

As these four challenges illustrate, there are major obstacles to establishing security in what has become the "center of gravity" of the United States. While increased connectivity and utilization of the Internet have led to economic growth and contributed to the dominance of our military, our increased dependence upon it leaves our nation at the mercy of its manifold vulnerabilities. Improving cybersecurity in the U.S. will require a multidisciplinary approach that encompasses public policy, law, and technology. Most important to this effort, however, will be a willingness on the part of policymakers and private sector leaders to have open and frank discussions about topics that have heretofore been ignored. The stakes are too great to let short-term ideological or geopolitical considerations prevent progress on securing the technology upon which our nation currently depends.

Eric Rosenbach leads the cybersecurity practice at Good Harbor Consulting and is an Adjunct Lecturer at the Belfer Center for Science and International Affairs at the Harvard Kennedy School, where he previously served as Executive Director. Mr. Rosenbach was most recently Managing Director of National Security with the Markle Foundation. Prior to this, he served as a professional staff member on the Senate Select Committee on Intelligence and as the national security advisor for Senator Chuck Hagel. Mr. Rosenbach's experience in the private sector includes serving as Chief Security Officer at World Online International, where he was responsible for all aspects of cybersecurity, privacy, and network operations. As a captain in the U.S. Army, Mr. Rosenbach was awarded the Meritorious Service Medal for his leadership of a communications intelligence unit. The Director of Central Intelligence named his unit the top intelligence organization in the U.S. military for two consecutive years. Mr. Rosenbach has co-authored several books on national security, military leadership, intelligence, and counterterrorism policy, and was a Fulbright Scholar in Eastern Europe. He earned a J.D. at Georgetown, an M.P.P. at the Harvard Kennedy School and a B.A. at Davidson College.

Robert E. Belk is a Strike-Lead qualified Naval aviator, who is currently a 2-year Politico-Military Fellow at the Harvard Kennedy School of Government (HKS). In addition to his operational aviation experience, LCDR Belk has served his country in various strategic capacities. Most recently, he served with Task Force-134 based in Baghdad, Iraq, presiding over hundreds of Multi-National Force Review Committee detainee review boards, including dozens for former high-ranking Ba'ath Party members. Under personal direction from the Deputy Commanding General for Detention Operations, LCDR Belk coordinated with the U.S. Marines of Multi-National Forces – West to improve the Coalition Forces' detainee operations in order to better support U.S. Counter-Insurgency strategy. Following the 2004 tsunami that devastated Indonesia, LCDR Belk became the Navy liaison from Carrier Air Wing Two to Combined Task Force – 536 to assist in integrating naval forces into the international relief efforts. As a Pol-Mil Fellow at HKS, LCDR Belk assists the Director of the National Security Program in developing and executing the South Asia Leadership Program. He is a member of the HKS Center for Public Leadership Student Advisory Board and director of its experiential leadership simulations. LCDR Belk is currently researching and developing policy recommendations for U.S. offensive cyber operations as part of his master's thesis. He has amassed over 2,500 flight hours in tactical jet and rotary-wing aircraft, and completed four carrier-based deployments to the Arabian Gulf, the Mediterranean Sea and the Western Pacific. LCDR Belk's awards include the Defense Meritorious Service Medal, two Navy Commendation Medals, Joint Service Achievement Medal, two Navy Achievement Medals, and various unit awards.

[1] http://www.whitehouse.gov/administration/eop/nsc/transnational-crime/threat

[2] Richard Clarke and Robert Knake, *Cyber War: The Next Threat to National Security and What to Do About It* (New York: HarperCollins, 2010), 89.

[3] The author recognizes that estimates of the costs of cybercrime and cyberespionage vary widely, but fall within this range.

[4] President Obama, "Remarks by the President on Securing our Nation's Cyber Infrastructure," Washington, D.C., May 29, 2009. Available at http://www.whitehouse.gov/the-press-office/remarks-president-securing-our-nations-cyber-infrastructure

[5] Ellen Nakashima, "List of Cyber-weapons Developed by Pentagon to Streamline Computer Warfare," *Washington Post*, May 31, 2011.

[6] Ibid.

[7] Based on the author's recent trip to Beijing to discuss cybersecurity with Chinese officials and researchers.

[8] Clarke and Knake, *Cyber War*, 50.

[9] Adam Segal, "Innovation, Espionage, and Chinese Technology Policy" (Statement before the House Foreign Affairs Subcommittee on Oversight and Investigations, Washington, D.C., April 15, 2011).

[10] "Chinese Hacked into Pentagon," *Financial Times*, September 3, 2007.

[11] Ibid.

"So tempting, easy, and apparently irresistible is the opportunity to steal, muck up, or generally just see if one can do something untoward across cultures and time zones that today hordes of empowered bad actors regularly overwhelm key security processes within critical complex socio-technical systems."

—CHRIS DEMCHAK

Resilience, Disruption, and a "Cyber Westphalia":
Options for National Security in a Cybered Conflict World[1]

Chris Demchak
Professor
U.S. Naval War College, Strategic Research Department

Caveat: All statements in this document are those of the author and cannot be attributed to the U.S. Government, the U.S. Navy, or the U.S. Naval War College

For two decades across Westernized democracies, national leaders, their corporate peers, and private citizens treated the enormously complex integrated networked system growing up underneath their feet as if it offered only free goods. Set by cyber-prophet visions of a new world village, the shiny aspects of vast unfettered data exchanges mesmerized most economic and political policymakers to think cyberspace would be unique in human history—beneficial, benign, and conflict free, or at least neutral.[2] The underlying presumption was that the elements of cyberspace could develop ever-increasing massive value in the central economic pathways of nations, and yet be free of human pathologies and conflict.[3] The consensus rules among the Cold War's dominant Western trading states guiding the international system were presumed to work the same way even if operating through a globally unfettered cyberspace.[4]

Exponential network and technological advances enabling witting and unwitting massive cybered transfers of wealth out of Western nations, however, have made cyberspace into a conflict space. Rather than the benign mutually governed international commerce arena envisioned by its early advocates, cyberspace has facilitated asymmetric strategic game-changing properties for the international system. Now public and private decision-makers in large and small Western democracies are relearning the lesson of history: Where great economic value emerges, so does great motivation for conflict, irrespective of rules established under previous conditions.[5] Because its globally critical properties allow the massive, near instantaneous, and often covert movement of knowledge assets[6] across national borders, cyberspace is changing economies and, accordingly, influencing the rise or fall of nations.[7]

Decision-makers of Western nations now face an extraordinary cognitive challenge in preserving the wellbeing of their nations. The global ubiquity, ease of use, pervasiveness, openness, and enormous complexity of cyberspace muddle the applicability of existing theories of conflict. From assumptions about likely avenues of crisis, values and threats from openness in international markets, and available state strategic buffers, the sea of variously capable, continuous, asymmetric threats from anywhere confound normal expectations. It is as if ancient, intelligence-poor Sparta were suddenly able to prepare its annual fall attack on Athens by spending the rest of the year remotely reading the intimate value, technology, and route details of every trade, and monitor every ship arrival or departure. The Spartans could then have reached into houses to change the data held by any Athenian merchants, their bankers, the leaders defending Athens, and even their suppliers overseas to ensure a poor resilience against the coming Spartan attack. Even if the relatively backward, non-trading warrior state of Sparta had not itself been willing to use this foreknowledge, Persia's emperor would have paid treasures for such advance revealing data in order to finally conquer Athenian Greece and control the Aegean for his own economic needs.[8] Long before the plague of 431 BCE decimated Athens, Sparta or Persia would have made Athens into a minor village, at best. Today, enemies accessing cyberspace have on hand the kinds of signals intelligence only wealthy states or superpowers could gather in the Cold War, and they are reaching for more in massive efforts to gather data, gain control of machines for future uses, and in some instances disable their distant surprised victims. The cognitive challenge is so profound that Westernized nations do not have the frameworks, metrics, situation awareness, processes, and science to fully understand the global socio-technical system and its inherent surprises that they have optimistically wrought.

Between the oft-confused decision-makers and the determined cyber-miners seeking advantage in droves, there is not an arms race, but a useable knowledge race. With the advantage currently to the attacker, the unexpecting defenders are now recognizing the cyber sands sinking around their centrally critical domains of economics, security, and social capital. Cyberspace has in the past two years been elevated from a support concern to a major national security concern in nations such as the U.S., the UK, Germany, and France. Each has issued at least one major strategy statement on defending against attacks in cyberspace. The 2011 U.S. National Military Strategy states under the "Deter and Defeat Aggression" principle that "A prosperous and interconnected world requires a stable and secure environment, the absence of territorial aggression or conflict between states, and *reliable access to resources and cyberspace for stable markets*" (emphasis added). (p.7) The new German cybersecurity strategy, issued early in 2011, states that "Ensuring cybersecurity has thus turned into a central challenge for the state, business and society both at national and international level." (p.2)[9] The new French cybersecurity strategy, issued in 2010, likens the national need for an effective response to attacks in cyberspace to a "new Thermopylae"[10] for the nation. (p.2, in French) The UK national security strategy of fall 2010 listed national cybersecurity as a 'tier 1' national concern. (p.11)[11]   — *Again in 2015*

Security professionals across these democratic, digitized states now must learn how to accurately conceptualize what is required to defend in, over, and through this currently unfettered man-made global substrate. Their choice of options, language, and connecting visions of national defense must implement the most adaptable set of strategies and institutions for a world of highly complex surprise generators not conveniently controlled by a peer or semi-peer state with whom to negotiate.

This essay argues that as the world's states most dependent on—and harmed by surprises in—cyberspace, Western nations need to recognize the rise of "cybered conflict," the change in the nature of national power to incorporate cybered aspects, and the need for a national security framework to identify and accommodate in advance the potential generators of nationally significant systemic surprise. To do so, what is required is a national strategy based on a combination of resilience and disruption capacities, a "security resilience" strategy and its associated institutional implementations. Using the security resilience approach, this essay outlines what is threatened by complex surprises from cyberspace, and what is meant by national "cyberpower" and "cybered conflict." It then explains how a strategy of resilience and disruption can accommodate four levels of complex socio and technological surprise. The essay briefly reviews what is needed to collectively incorporate security surprise considerations in organizational and technological designs across four kinds of surprise, and the need to recognize and technologically guide the rise of protective borders in a new cyber Westphalia process. Finally, the essay evaluates the most explicit national statement of U.S. response to cyberspace's multi-layered threats— the new Department of Defense Strategy for Operating in Cyberspace (DSOC)— in terms of having considerable elements appropriate for a nascent resilience strategy. It argues that the DSOC guidance could be improved in three specific areas: organizational and technological design learning about surprise-resilient processes of globally blended socio-technical systems, the implications of the rising cybered Westphalian process on the changing topology of cyberspace, and the integration demands of a disruption strategy specifically aimed at the most skilled bad actors able to surprise most resilience scheme measures at their will.

As the topology of cyberspace changes and cybered conflict more clearly emerges, more options for national cyber defense need to be developed and continuously tested. If done well enough by the states most at risk and most able to change the cyberspace substrate they themselves built, these efforts will spur technological advances for secure design of basic systems, retain the necessary interoperability for commerce in a mobile IPv6[12] world, and increase privacy-sensitive knowledge development for accurate national security situation awareness and informed policy choices. In short, it is possible for states to understand and accommodate cybered complexity well enough to inch back to the world promised by the cyber prophets twenty years ago. In this possible future international system, states assume responsibility for traffic transiting in or out of their national networks and for restructuring the underlying security of the technological base. This cybered world is better structured for rules of honest commerce and minimal opportunistic malicious activity. It is likely to offer fewer ways for complex systems surprise to be used by jurisdictionally unreachable actors to harm others, and more dynamic technological advances that are designed and implemented to incorporate security for the individual and overall system as well as general economic wellbeing.

Complexity, Surprise, and Cognitive Challenges for National Defense

Cyberspace poses an enormous globally integrating socio-technical system of such unprecedented complexity that the lessons of history and science can offer only partial insights. Defense options, strategists, and organizations must deal with enormous sources of uncertainty whose systems-threatening surprises would be hard to accommodate in their own right, even without aggressive actors. These include the ill-designed complexity of the fundamentally insecure technological layer now forming the main nationally critical cyber substrate, the associated potentially devastating surprises inherent in large-scale complex socio-technical systems (LTS) formed by the substrate and the dependent societies, and the extraordinary volume and persistence of malicious actors reaching in from anywhere around the world to covertly steal, control, or damage. Together they sum to levels that overwhelm the analytical and response capacities of national leaders.

In this emergent reality, kinetic or clearly defined indicators of the onset of state-state 'war' are unlikely. Rather, conflicts will be "cybered." They may start in networks, but will not stay there. Cybered conflicts are those nationally significant aggressive and disruptive conflicts for which seminal events determining the outcome could not have occurred without 'cyber' (meaning networked technologies) mechanisms at critical junctures in the determining course of events. The ubiquitous, easily accessed nature of a turbulent, rapid, and massively 'noisy' cyberspace and its reach into even the homes of distant potential targets changes the way conflict begins and even its tools. Such struggles can begin long before all participants realize that they are the victims of long-running, largely covert or deceptive, existentially significant cybered operations.   — Shoud & Lowther 'Air Force Strategic Vision 2020-2030' p. 10

Compared to more legally recognized forms of war, cybered conflict is more holistic. The nature of the global cyber substrate is so extensive that a bad actor far from any jurisdictional control can readily choose the scale, proximity, and precision of any single or series of attacks. Unlike most of history, a bad actor can organize an adversary organization at any scale, from 5 to 5,000 participants; collect knowledge and reach out to strike at any distance, from 5 to 5,000 kilometers; and choose tools with any precision to disrupt or control victims, from 5 to 5 million nodes on the global internet. Such choices are historically unprecedented save for superpowers, making the current propagation range and potential effects of persistent and adaptive cybered attack close to complexly incalculable and an extraordinary cognitive challenge. When in full flower, cybered conflict will be:

- Broader (in terms of both wide propagation radii and effects);

- Longer in overall duration;

- Indeterminate in its signals, beginnings, and end;

- Stealthier as an efficiency measure for the attacker;

- Cheaper in knowledge burdens, resources, and loss of buffers for the attacker than for the defender;

- Opaque in the multiplicity and volume of possible signals of hostile action;

- Inclusive across the numbers and motivations of combatants able to be involved;

- Countering the resilience of defender rather than directly confronting; and
- Combinatorial in the vectors of simultaneous attacks possible across internal national sectors.

For the defender, these attributes are cognitively exceptional and challenge the established indicators and responsibility allocations of the modern Westernized nation. How can one know if one is under attack if none of the attributes advantaging attackers require overt behaviors by the large number of persistent, clever actors hiding in far-flung unsympathetic or incompetent states unable to control their own cyber jurisdictions? With sufficient access to the global cyberspace, these actors may readily obtain massive amounts of what used to be expensive signals intelligence, to include the likely sources of disabling surprise found in normal systems. With such knowledge, actors from adversary states to proxy or opportunistic free agents are able to push at any chosen tempo a myriad of many-against-one games with significant nationwide security importance to conditions that nearly reach but do not provide the obvious markers of Cold War crisis or kinetic trigger thresholds.

In a cybered world, the cognitive difficulties of complexity are overwhelming for defenders. The uncertainties directly challenge established and historically simplifying terms of conflict mitigation and conduct of conflict. In an openly integrated cybered world, defense and offense distinctions blend at the technology level. If some adversary has infected critical supply chains of key computer components across a nation's telecommunications networks, does that constitute aggression, espionage, or crime? What if the purpose of the infection is unknowable because the programmed trigger is undetectable, and the businesses or owners of the infected computer nodes refuse to pay for its removal without being forced by regulation or persuaded by a national demonstration of the harm? If the parts are literally everywhere across critical nodes and if there is a trigger buried in the infected software or hardware, the time to respond across the whole nation will be effectively zero, crippling the whole state in massive ways at a moment's notice. Prospective harm is difficult enough to argue in normal security circles; potential harm of this magnitude has proven exceptionally hard to establish.[13] What if one cannot wait for the big hit that proves the vulnerabilities because adverse individual consequences will today cumulate in minutes to hours to produce large systemic impacts?

Furthermore, the normal surprises of complex systems mean these rippling effects can be much more severe and out of control than planned by initiating adversaries. The effects in a globally integrated world easily ripple through interdependent couplings across many states at once. The Stuxnet worm spread widely in 2009-2010. All it seemed to do initially was to infect, open a secret backdoor, and harmlessly wait for instructions. There are so many such odd infections that the world of computer security has become inured to bits of code unless the bit demonstrates malicious behavior. [14] Stuxnet had a target; the specific basic controlling logic program used by Iranian nuclear reactors. Had it had a trigger to harm much more broadly—perhaps triggered by a certain date or event—however, it would have been near impossible to stop large portions of the cyberspace nodes it had infected from being disrupted. [15]

Making the cognitive challenges worse are the multiple ways threats of global cyberspace challenge the established political and institutional divisions of sovereignty between international and national responsibilities underpinning the modern democratic state. Within states, the domestic allocations of knowledge and responsibilities concerning cyberspace as the proper concern of national security, military institutions, homeland emergency or police entities, private commercial sectors, or individuals in their own communities and home systems are under stress. Signals recognized and policies enacted by one set of policymakers can be disdained or ignored by others whose sectoral responsibilities are not yet overtly or severely affected. The opaqueness of the wide range of covert yet hostile acts possible across cyberspace make escalating cyberconflict obscure and hard to recognize, especially since the huge volume of bad actor activity associated with rampant cybercrime easily obscures the efforts of more persistent adversaries.

Preparing the cybered battlefield has never been so cognitively easy; recognizing those preparations and neutralizing them technologically, societally, institutionally, and strategically has never been so cognitively difficult. In 432 BCE, Pericles of Greece could prepare for the inevitable Spartan attack precisely because he knew when, how, and with what Sparta would attack, even if not precisely where the army would come each year. He could, and did, manipulate an unwilling Athenian assembly to pay for a long, protected wall connecting the city to its critical port, knowing that with this access a Spartan land siege could never take down his city. He needed only three days' notice to bring in all of Athens' farmers—an emergency response that was legally and physically enforced—and thus spies able to give three days' notice provided Athens critical resilience which only a plague would ultimately destroy. [16] Today's would-be Pericles cannot even say if a Stuxnet-like worm is already present in most of the key nodes across the modern integrated national banking or airline systems, and yet must somehow prepare the society for that possibility while further preventing any future infections and external control.

At the end of the day, the cyberpower of a nation will depend on its ability to deal directly with complexity, surprise, and the multiplicative effects of active human cybered threats in an integrated world in order to maintain national economic wellbeing, social capital resources, technological integrity, and strategic power. As argued below, national power in a cybered world will depend on two broad capacities to deal with this cognitive stress: the ability to be nationally, systemically resilient when cybered surprise happens and the ability to forestall surprise by disrupting sources legally, legitimately, and effectively.

Linking Complexity, Surprise, and Threats to Security Resilience Action Options

The goal of a security resilience strategy is to reduce this cognitive challenge for national leaders and institutions by neutralizing many lower-level generators of complex systems surprise and allowing focus on particularly complex threats. The object is primarily to have resilience across the nation dependent on its digital substrate; that is, to have sufficiently accurate, timely foreknowledge and well-tested contingency elements across quickly and responsibly recomposable pieces already in place to dynamically accommodate systemic surprise when and where it happens. From the incoming, continuously reproducing hordes of bad actor operations to the unwitting or ignorant actions of individuals or jostling organizational entities with self-interested conflicting stakes in the security of their own cybered knowledge assets and exchanges, the goal is to anticipate the form or frequency of surprises with serious propagation range and effects and to prepare to curtail their spread and mitigate their effects immediately.[17] Resilience in complex socio-technical systems is not achievable by copying static best practices in an overarching standard design of technology, organization, or policy. Rather, it requires a combination of dynamic actions that staunch surprise-generating processes in advance and mitigate threats as they emerge.

For national defense decision-makers, the requirements for resilience are not so much how to array full spectrum offense and defense contingencies in war or peace time, but rather how to recognize and then organize collective sense-making and innovation to accommodate the surprise threats emergent throughout the globe's complex socio-technical cybered systems. To do so, they must deal simultaneously with four overlapping layers of increasing complexity across the global cyberspace, each generating ever-higher levels of multi-source threats: (1) the surprise generators inherent in largescale complex socio-technical systems, (2) complex critical societal socio-technical systems, (3) cybered bad actors throughout global systems, and (4) the exceptionally skilled persistent "wicked" cybered bad actors. For each, scientific and experiential literatures outline a set of surprises and adaptive responses, scalable from the enterprise to the national levels, as described in Table 1 below.

First, the surprises of large-scale complex systems require structured and adaptive resilience based on redundancy, slack, and effective discovery trial-and-error learning (DTEL) in people and machines.[18] In cyberspace, the surprises are fundamental incompatibilities when the precision needs of one set of processes are not met in designs, operations, or evolutions. In 2011, the antivirus firm McAfee issued what is now considered a standard update. Unfortunately, the update was mistakenly coded to see a small critical file in the Windows XP operating system (a .dll file) as malicious. When the update was installed, the "malicious" file was disabled by the McAfee program. When the machine was turned off or the user logged off, Windows could not open up again because its internal software could no longer find that file. Computers across thousands of locations, including whole institutions in the U.S. military, had to be physically rebooted in a safer mode and older, uncrippled versions of Windows restored one by one, a days-long process during which many could not work.[19]

This kind of surprise and its effects are not uncommon; they are the "normal" accidents of such complex systems. Had the underlying technological design not needed to allow such incredible access and power to an external AV program, the surprise would not have happened. Had users or system administrators had indicators that such a critical file was being quarantined, even if by a trusted program, again the costs and losses would not have occurred. Had the system being changed or the program doing the changing been open to informed users selecting what files to change protectively, then the disruption would not have occurred as far or as deeply. Had McAfee not been addressing a standardized operating system found in roughly 80 to 90 percent of the computers of the world, it is more likely other systems in the same organization could have redirected to keep the operations alive while the update miscoding was mitigated. Resilience in place would have had elements of all these recommendations: baseline secure underlying technology, sensors telling well-informed users of a critical change that they could recognize as a bad idea, the ability to refuse such a change incrementally while allowing other critical updates, and the redundancy of knowledge held in easy substitutes using unchanged reliable systems. Responses at this level ensure redundancy on the spot in the knowledge needed, build in slack in time to forestall bad effects while the missing knowledge is recognized and acquired, and continuously develop discovery trial-and-error learning that guides and adapts the redundancy and slack decisions.[20]

Second, for the surprises emerging from massive socio-technical complexity affecting critical national systems, larger and more inter-organizational forms and continuous adaptation of resilience is necessary. When nationally significant critical infrastructure systems are at stake, no one small set of decisions is adequate to accommodate all the variations of possible surprise. Resilience for these systems builds on the routine responses to surprise by deliberately incorporating widely collective sense-making and rapid mitigating improvised action equivalent to the potential propagation range and disruptive effects of widespread urgent conditions.[21] Resilience here is inherently inter-institutional. Trust-building across communities, sectors, and all relevant organized entities is critical along with sensor systems, redundancy of knowledge, and slack in time in order to have access to the knowledge pools of all the critical players in advance, during, and after the urgent disabling event. For example, in 2007, Estonia's entire heavily cybered national government and financial system found itself under a massive disabling denial of service attack coming in seemingly from all over the world. Fortunately Estonia has a small, particularly patriotic population in which a handful of computer science experts in and associated with the Estonian government were able to work together on the basis of prior friendships and trust to literally take control and defensively shut off the government to save the rest of the system.[22] By serendipity, Estonia had in place the basic elements of complex critical infrastructure cyber resilience due to the small groups of decision-makers involved and their ability to know and trust each other well enough to immediately collaborate. Scaling that advantage up to the size of the United States, however, requires continuous, advance, deliberate, and multiple efforts at developing the necessary collective sense-making, trust in people, technological options, and sensor validity to know how to respond to urgent cascading surprises.

Third, individuals with access to the Internet often lose their social constraints to act maliciously and vastly multiply the normal surprises of complex systems and the more nationally dangerous urgent events in critical infrastructure. So tempting, easy, and apparently irresistible is the opportunity to steal, muck up, or generally just see if one can do something untoward across cultures and time zones that today hordes of empowered bad actors regularly overwhelm key security processes within critical complex socio-technical systems. While it seems normal to speak of the U.S. Department of Defense being hit by millions of attack attempts daily, a vast number of nonmilitary institutions are also daily losing money, the integrity of their files, the reliability of their programs, and the security of their investments in social capital, intellectual property, and even personal security. The losses have, until recently, been seen as mere vandalism, or at worst as cybercrime in information theft; events best left to domestic institutions and not worthy of national security concern. However, the magnitude of the bad actor successes is exponentially increasing. For example, in 2010 the top six corporations in the U.S. lost $130 billion dollars to cybercrime.[23]

The successes of these huge floods of bad actors matter for the cognitive challenges and resilience responses of national security leaders because they do not simply drain a nation of knowledge resources or leave secret back doors behind for future thefts. The international cybercriminal community functions as a huge global laboratory constantly innovating and demonstrating new, low cost, effective techniques for covert rapid access, extraction, and control into networks, critical functions, and essential knowledge bases. Those bad actors with larger resources, especially states, easily adopt the innovations of cybercrime, adapting them as needed, and then use the enormous volume of criminal attacks as noisy cover for the more particularly malicious operations. The cognitive challenge is currently so great that resilience in terms of post-attack mitigation is simply insufficient and more attention to pushing the possible surprises further from their targets is necessary. As bad actors multiply the confusion, more slack is desperately needed in buying time for responses and for channeling the propagation effects. In the language of traditional organizational information theory, more gateways and other "air gaps" in terms of automated and human interfaces to systems are necessary to reduce the input flow of uncertainties from cyberspace and to curb or channel the way these systems interact with their environment.[24] Bad actors are forcing nations, in effect, to develop the building blocks of borders in cyberspace in order to curtail the floods overwhelming internal systems and to buy time to recognize and collectively develop new responses to those surprises that inevitably will make it through anyway.

Fourth, and finally, within the mass of geographically-dispersed bad actors meddling in critical systems and spawning unmanageable complexity are a smaller set of "wicked actors"[25] who, like the wicked problems of mathematics for which they are named, are significantly increasing the cognitive challenges to national security leaders.[26] This is a smaller but exceptionally threatening set of deliberate adversaries who require more than just the routine mechanisms of resilience, because their skills defeat most normal responses. For example, while the social engineering tactic used by the attacker getting inside the RSA corporation in 2011 was well known across the cybercrime community, the skill and prior intelligence required to know what was sought and how to use it suggested the perpetrators were far more skilled and dedicated than the vast majority of bad actors. A string of successes exploiting the encrypted tokens used by major defense and commercial corporations resulted from the initial attack, forcing major institutions to literally take their networks offline to repair the vulnerability and losses.

The presence of such highly skilled wicked actors complicates the cognitive challenge in ways even the variability of the flood of routine bad actors cannot. The resilience responses of redundancy, slack, DTEL, collective sense-making, and even slack in gateways or borders are insufficient. For these actors, effective response strategies must combine lessons of the resilience literature with the field of study on war, security at all levels, state and hostage negotiations, economic incentives, cultural imperatives and legitimizers, and comparative human sociopathic tendencies. For them, the national responses require disruption capacities able to derail bad activities by intervening in their OODA loops[27], business models, or motivations before they achieve access internally to an institution, network, or state.

Together, these efforts change national security in cyberspace from either a civilian or a military endeavor involving crime or war to an *integrated security-resilience strategy* combining these four sets of responses. Successful adaptation to complex environments requires combined systemic resilience to surprise at the normal levels of complexity, and then more intensely when exceptionally large complex systems are also directly critical to national systems. Proliferation of bad actors globally requires additional disruption operations against particularly effective state and non-state wicked attackers. Table 1 outlines the layers of complex surprise generators and the resilience responses dictated by the addition of each layer.

Implementing the cyber resilience actions outlined in Table 1 requires capacities at each level adapted to the surprises of that level. These include:

1. Layered human and machine sensor sets and local knowledge development processes and tools so that the redundancy of knowledge, slack in time for innovative responses, and sufficient preparatory discovery or trial-and-error learning (TEL) will be in place precisely when and where needed.

2. Widely dispersed adaptable tools for immediate sense-making and action so that organizations under urgent conditions can effectively and collectively know what has happened and what could happen, can collectively consult on the spot with all other critical actors, and can then act rapidly and accurately with real-time feedback for adaptive collective corrections during the process.

Table 1. Requirements for Resilience and Surprise Management at
Four Layers of Threat and Complexity

MultiSource Threat Categories in Increasing Uncertainty and Surprise Potential	Cybered Resilience Action Requirements (including Disruption Supplement)
Complexity in Largescale Socio-Technical Systems, LTSs (basic "normal accident" and cascading surprise-prone large cybered organizations)	1. Redundancy (of knowledge) 2. Slack (in time to respond) 3. Organizational Discovery Trial-and-Error Learning (DTEL)
(all above) plus **Criticality for Nation** CIP, Critical Infrastructure (protected status), High Reliability Industry, or Operationally Engaged Military	(all above) plus 4. Collective Sensemaking 5. Rapid Accurate Mitigation, Improvisation, and Adaptation Action 6. Frequent Whole System Practice for Extreme Events under Urgent Conditions
(all above) plus **high volume Bad Actors** (average to good skills, ubiquitous from script kiddies to vast majority of botnet masters, volunteer anarcho-hacktivists and less-skilled nation-states)	(all above) plus 7. Enforceable Cyber Hygiene 8. Underlying Technology-Secure Design Transformation 9. Comprehensive Multi-Organizational/Layer Learning for Systemic Generative Innovation 10. Stratified Two-Way Flow Sensors/Tagged-linked Interoperable Policy-guided Gateways ("National Cyber Borders")
(all above) plus **Wicked Actors** (high threat persistent motivations, exquisite skills, ability to organize, access/evasion expertise, or wide deep harm propagation potential)	(all above) plus 11. Extensive Wicked Actor(s) OODA Loop/Business Model/Motivation Knowledge Collection and Development 12. Selected Controlled Disruption under Protective Principle of International Law 13. Collective Understandings/Undertakings with Like-minded Cyber Responsible States

3. Structured organizational knowledge flow and advanced technological designs regularizing knowledge transits into and out of human and machine systems in order to regulate the volume and quality of inputs likely to impose surprises to overwhelm the internal processing capacities built by the socio-technical sensor sets and knowledge development and the collective sense-making and accurate action tools. (For one idea on how this might be done in the nearer term at national levels, see Mallery's borders technology thought experiment in Appendix A.)

4. Specialized disruption capacities to ease the pressure on numbers one, two, and three, above, and reach outside to selectively alter the motivations (legitimacy, need, confidence)[28], business model, and OODA loops of wicked actor groups under highly controlled conditions and in accordance with established laws and permissions.

With these capacities, a nation has a much better set of options for defending their wellbeing in a heavily cybered world. All options are exquisitely knowledge intensive, and addressing their feasibilities and consequences is essential for a discussion of resilience at the national level. Two aspects of national security-cum-resilience, however, need further discussion, not only because of the political issues surrounding each but also because of the challenging practical implications. The two issues are the rise of borders in cyberspace and the use of disruption as a part of a national security resilience strategy.

Rising Cybered Westphalia as a Resilience Option for States Seeking Certainty

Gateways in cyberspace are widely being instituted today not because decision-makers intrinsically desire them, but because they are an available and natural system-level response to overwhelming cognitive uncertainty about harmful surprise from the external environment. The goal is to ease the pressure on the interior's ability to withstand disabling surprises. The lessons from classical information theory and living systems theory endorse insulating internal processes from damage or subversion by the environment. One way to do this is by hardening the internal processes to make them impervious to disruption and able to process more volume effectively even under urgent conditions.[29] For national resilience, hardening includes better system-wide hygiene, knowledge redundancy, organizational trial-and-error learning, higher assurance in security across technologies, and transformation of the underlying technological layer.

The second way of insulating internal processes is to somehow constrain the disruptive inflow.[30] In cyberspace, constraining inflows means labeling, stratifying, managing, inspecting, disinfecting or otherwise making decisions at frontier gateways about network traffic entering or exiting. Historically, this method was more common than increasing internal processing power. From moats to minefields to drawbridges, humans operating under uncertainty have tried to push the threat as far from themselves as they could so that they had more warning time to process a possible solution when surprised.

Today, the current flood of assaults along network attack vectors is so vast that simply increasing internal processing has not proven to be enough to curb an attack. Now both enterprises and states are nearly instinctively attempting to better curb the inflows containing surprises or at least the possible propagation range of overall effects once inside. From firewalls to cyber demilitarized zones and other tools that enable decisions over what is permitted to enter or change critical internal systems, national institutions are slowly accreting the building blocks of a strategic buffer in cyberspace; that is, a national cyber border against accidental, imposed, or enhanced surprises flowing in or out of their traditional boundaries of sovereignty.[31] The topology of cyberspace is changing in this emerging cybered Westphalian process. Already one can identify three distinct models of responses to national insecurity in cyberspace: the Chinese, or "all points" model; the "key firm" model of major European states; and the "cyber command" model of the United States and now other states, including South Korea.[32] New models are emerging as well—notably in smaller nations with specialized digital circumstances, such as Estonia or Australia— as nations seek to control the uncertainty flooding their networks. This topological change will inevitably alter distribution and manifestations of national power, and ultimately the character of each nation's future quality of life.[33]

If the emerging cyber-Westphalian process is to achieve all those things, however, the socio-technical designs of these borders need to be guided by deliberate, informed and thoughtful strategy more than by serendipity across the mass of individual institutions of a nation. John Mallery has devised a draft technical framework based on traffic tainting for implementing digital borders at network speeds.[34] This timely "thought experiment" offers a mechanism to incentivize important actors, including governments, ISPs and other large organizations, to accept responsibility for network traffic originating from their zones of administrative control. It accomplishes this by irrefutably identifying transit traffic and, by induction, assigning origination to an actor who can police any malicious activity; it can be implemented top-down by states or bottom-up by ISPs and enterprises. Whether a cyber border is operated by a nation's uniformed agents or by regulated coordination across internal and institutionally maintained gateways, its success will require considerable strategic consensus and extensive collective coordination to be legitimate, orderly, interoperable, and ensuring privacy.

If the rising borders are effective, there can be advantages in the transitional development of credible national cyber sovereignty. Recognized cyber borders could indeed reduce the flood of maliciousness buried in internet packets, induce technological innovation, reveal information in broad trends unobtainable now, provide cognitive simplification and closure, and finally, work to make states responsible for what their inhabitants do to the wider cyber ecosystem. The need for technologies secure in their design and implementation to sustain digital borders will stimulate demand for more secure internal and international networks than the original designers of the Internet anticipated or deemed necessary. The search for built-in security is likely to incentivize economic investments and innovations within and among Western countries as well as around the world. As the borders rise, economic exchanges do not need to be hampered—gateways do not automatically make interoperability across a net; the flow depends on the choices of the designers and operators. Effective cybered borders require more than centralizing traffic gateways, such as the three portals to the global Internet operated by China.[35]

A well-designed boundary of national cybered sovereignty allocates risk and responsibility across actors able to respond. It can enable nations to decide what risks are allowed to flow in to a much greater extent than possible today. Recognized cyber borders highlight state level accountability for their cyber emissions as well. Clarity here helps international development of good neighbor norms, international help for those states unable to police their cyber presence, and collective sanctions for poor state behavior in cyberspace. Effective "rules of the road" would have a better chance of being successfully implemented[36] if realistic mechanisms for verification of irresponsible state behavior were available. Furthermore, a host of well-established theories and understandings developed for the non-cybered, physical world of national sovereignty would be more applicable to the speed, range, and complexity of cyberspace and the sovereign responsibilities of a global community of more clearly defined cybered states.

Disruption as Part of a Resilience Strategy

For the majority of possible surprises and malicious actions by bad actors, resilience is the primary and most effective response. But an open cybered nation must have the disruption capacities to allow it to focus on what is needed when wicked actors are involved. These exceptionally persistent, skilled, lucky, organized, or closely coordinated actors in the top 5 percent of the wider hacker community can routinely overwhelm normal resilience mechanism measures. Their behaviors demonstrate malevolence or nationally significant malfeasance tied to exquisite levels of on-call IT manipulation skills, effective organizational capacity (including resources and evasion), exceptionally talented access and remote control/extraction activities, and/ or highly-likely wide harm propagation success. Wicked actors are a small set, but once they get inside, they generally cannot be detected or stopped in time to prevent an attack.[37] Disruption operations aim to ease the pressure on the rest of the nation's resilience mechanisms before these actors succeed. The objective is to directly or indirectly increase the wicked actors' perception or experience of obstacles before they are engaged in an operation.

The knowledge challenges to effective disruption are not small. A nation cannot rely solely on disruption to protect itself, given the flood of active, independent, covertly operating bad actors with unclear incentives, individually creating obscured patterns of cyber noise with erratic behaviors used as cover by wicked actors. The set of wicked actors to pursue must be pruned and the costs in knowledge collection and development terms or in the intricacy of careful, successful disruption operations are not reasonably affordable if the target set is huge.[38] Today, attribution is practically and legally possible only if the target set is small enough that bad behaviors can be accurately charted and the multiple jurisdictional permissions obtained for each case. Large amounts of data, not always of a military nature, will be necessary to clearly identify those engaged in long-running covert, extensive cybered operations against defenders' knowledge assets. Even with a much smaller set of targets, however, disruption operations will need to finely focus on the wicked actors who pose the biggest threats not susceptible to other dampening means found in national security resilience mechanisms such as hygiene, borders, more secure baseline technologies, and broadly promoted norms.

Furthermore, care must be taken in identification and disruption operations because the indicators of cybered conflict will be subtle amidst a great deal of unrelated cyber noise. Cybered conflict begins unannounced because an active, declared attack is usually inefficient for most objectives of obtaining knowledge, control, or backdoor access for future operations. By remaining hard to reach, and given the low costs of operating globally through cyberspace today, wicked actors have the capacity to conduct a low level but still enormously profitable or productive cybered conflict for a very long time. Their patterns of action can include bleeding off particularly valuable knowledge assets critical to the comparative economic or power advantages of the defending nation by inserting controls in key systems remotely or through corrupted supply chains, or by selectively directly enfeebling. Any set of actions could constitute a preparatory phase, a main avenue of attack, a central campaign element, a large-scale deception and espionage operation, an episodic enabler, a foregone set of activities, or all of these at different times in a long-term cybered conflict. Wicked actors readily adopt and refine the techniques of cybercrime used by the mass of cybercriminals operating globally, thus making the defender's data collection and development challenge even more difficult.

In fact, due to the ability of particularly skilled actors to hide for long periods, what is discussed today as openly aggressive "cyberwar"[39] is likely in reality to be the later stages of a cybered conflict. Particularly persistent and skilled adversaries will attempt to ensure future benefits by covertly pursuing "counter-resilience" campaigns to assure no effective defense is possible if the attacks become known to the defender. An adversary state might be engaged in this counter-resilience behavior to achieve short winning conflicts or simply to lay the groundwork for later, preferably bloodless concessions, making attribution highly political as well as operationally demanding for disruption operations. The equivalent exists in cybercrime, where a botnet master actually cleans off other botnet infections from the computer they are infecting, inoculates the infected computer against competitors, and changes the infected computer's antivirus so that it does not alert on the presence or activity of the botnet software when the owner is using the machine. Then the botnet master can use the computer at will without being revealed even when the unwitting owner is online.[40] Even with small numbers of wicked actors, disruption techniques need to be able to focus through the miasma of data across economic enterprises, infrastructure systems, military/intelligence/police structures, or other critical national institutional and technical means in order to see what activities do and do not constitute part of a wicked actor's counter-resilience campaign.

While the mass of data is daunting, the good news is that it becomes possible to disrupt these actors because their preferred tool—cyberspace—logs actions, records results, and responds to programming, systems administrators, and technological design constraints. Thus, patterns of wicked actor behavior are discoverable over time across servers and network exchanges with a determined, sophisticated, expensive, and usually time- or manpower-consuming, effort that ranges widely across cyberspace's signals. Effective disruption operations can then use this data to make it harder for wicked actors to operate, forcing them to accommodate more challenging work factors in their operations. Ideally, with sufficient probable cause and international jurisdictional permissions, a disruption operation is precise enough to disturb the wicked adversary's OODA loops in real time, their business cycle model, or, over the longer term, their underlying action equation (legitimacy, need, confidence).[41]

Disruption has the potential to slow wicked actors down. If it is more difficult to operate, adversary actors will need more time to lay in greater amounts of covert leverage than they would normally in order to maintain the economic resources for future exchanges. Afraid of being targets of successful disruption, wicked actors will have to work harder to be stealthy and to delay the defending communities' recognition of their cybered losses and covert controls or backdoors. When resilience rises, the successes of random bad actors become more a matter of accident than skill. The activities of wicked attackers, then, will stand out, signaling a malicious actor of deeper interest to both defending and hosting state authorities. When the global cyber noise is no longer a good cover—and may, instead, turn into a spotlight—a covert plan might be revealed or destroyed by the actions of an eager set of unassociated fellow travelers of lower skill who pile on to score a few free, unpunishable points.[42]

Disruption can help force wicked actors to accept more personal risk in their 'work factors'[43] calculations even if the defending nation's disruption capacities are imperfect in advance—that is, the nation has limited cyberpower overall. Among the three major advantages cyberspace offers to bad actors (scale, proximity, precision), disruption operations in the short- and medium-term are better focused on confidence linked to proximity, making the risk personal to the individual, groups, or even states engaged in this behavior. In a more nationally resilient global community of nations, distance will no longer be an advantage. Even the most capable attacker is likely to have to take risks not necessary today, perhaps even personally entering the protected nations and making the physical identification for disruption purposes easier.

Furthermore, with a serious possibility of being identified, the wicked actor operating for a peer state will have to work harder to avoid being revealed and destroying the leverage of a long-running covert cybered conflict campaign before it is in the attacker's advantage to go public. Failure to remain stealthy could have unintended consequences that include possibly destabilizing an otherwise only slowly intensifying cybered conflict. The outraged defending nation and its allies could miscalculate and escalate the situation well beyond the conflict levels intended by the state employing the wicked actors.[44]

Disruption can also impose a follow-on cost on successful wicked actors unavailable in the national resilience mechanisms. Once the routine bad actor has been neutralized, the direct interaction is over. For a particularly successful, persistent bad actor, stolen data can continue to reap benefits long after the data is exfiltrated, the backdoor discovered, or the original exploit patched. Having disruption capacities allows national authorities to alter the continuing flow of benefits by intervening in the parts of their business model tied to selling it, using the covertly inserted controls, or developing greater wicked actor confidence in future attacks. A disruption success can be widely or discretely publicized as a discouraging signal to a wider community of both wicked and routine bad actors. Successful disruption by any national institution or major enterprise will be discussed in hacker forums read by most bad actors from state levels to script kiddies, and thereby serve as signaling corollary benefits beyond diminishing the benefits of a successful assault.[45]

At the end of the day, disruption capabilities are no panacea in national cybered security resilience, but they do buy time for the defending nation by pushing the most serious deliberate sources of surprise—the wicked actors—further in resources, time, personal risk, and complexity from successful operations. Disruption should never been seen as an easily conceived secret "offense," cognitively simplifying the response to the threats of the nation. Rather, effective disruption capabilities are more like expensive long-term cyber SWAT teams. They are used sparingly because ideally they require careful double-loop learning, long-term informed target selection and timing analysis, exquisite abilities to 'play through' the downstream and lateral implications, and a policy of willingness to let resilience take the heat if the disruption plan is weak. Disruption operations are really only useful for the fourth level of complexity.[46] To be effective, disruption must be built on a foundation of normal resilience mechanisms of a nation dealing with the first three levels of complex surprise generators. A nation that relies solely on only one narrow form of resilience from hygiene to borders to disruption can easily find its knowledge and security options limited.

DOD Strategy for Cyberspace: A Nascent Resilience Document

As an initial strategic document signaling a large organizational transformation by a major national institution, the new U.S. Department of Defense *Strategy for Operating in Cyberspace (DSOC)* provides considerable promise as a nascent resilience strategy. While some have argued that it is insufficiently precise in its recommendations,[47] the DSOC shows the beginnings of a reorientation toward accommodating complex surprise rather than focusing solely on prevailing against identified military combatants. As shown in Table 2 below, elements of the DSOC correspond to the resilience strategy concepts outlined in Table 1. While the DSOC cannot be expected to address all conceivable considerations in detail, the match between the resilience requirements and the aspirational statements of the DSOC is very encouraging.

Nevertheless, the document has shortcomings. The scattered pattern of the resilience related statements across the initiatives reflects the lack of a self-conscious resilience orientation at this stage. To move toward a resilience focus while defocusing legacy notions of offense-defense operations, a major strategy document must address the implications of three areas on the strategic guidance: the demands of organizational learning under surprise, the effects of borders in cyberspace on conflict training and guidance, and the implications of a widespread need for and use of disruption capacities.

First, organizational learning needs more structured prominence, involving not only general promises to better train, practice, and innovate but also better expression of the integration of process, technologies, and institutional changes required for the first two layers of resilience. Given the complexities faced today and the consequences of failure, organizational learning cannot be a random bag of well-intentioned efforts; it will require much more integrated, collective knowledge development than the DSOC implies. Organizational structures will have to adapt. Appendix B discusses one option, the Atrium model, as a scalable organizational model of cyber-enabled tacit knowledge acquisition and development for widespread organizational learning to accommodate surprise within and across institutions.

Second, as borders rise in cyberspace, the conflict space will change with the new topology, from new developments in technology to sustain this evolution to greater difficulty for routine bad actors to cross into the nation's cybered space. This trend has considerable implications for the DSOC's resilience choices in technologies and in disruption policies, but it is not currently addressed in the document. Futures tend to be path dependent; channeled by events and decisions in the past. The choices the DSOC encourages today need to be able to accommodate a future in which states assume sovereignty and take responsibility for their cyber territory. If degraded, the DOD's technology strategy to secure its own cyber perimeters, defend in depth, engage in active defense (a subset of disruption), and operate will drive innovation in the wider IT capital goods industry and related sectors. DOD's own specific advanced requirements and the magnitude of its catalytic buying power have considerable knock-on effects. For example, DOD may require more technological resilience components be manufactured "safely" inside national borders to minimize externally tainted supply chains and impose particular standards for a "secure" cloud architecture, both choices which may or may not be congruent with the rise of cybered borders and the technological innovations necessary to make them effective.

Third, the role of disruption receives exceptionally limited coverage in the DSOC. Due to the necessity of having some response to wicked actors, it is likely that disruption operations will occur in any case. However, with no formal discussion of the implications, benefits, and limitations of such capacities, the systemic effects of disruption's operational choices could be unwittingly counterproductive. For example, resilience guidance will need to be altered if disruption is taken out of the toolkit available for the fourth and hardest set of surprise generators. Unfettered use of disruption can result in abuse or misuse, perhaps leading to kinetic action or inducing a global declaration that disruption is an act of war and restricted only to use during active conflicts likely to involve kinetic action. With little formal guidance on the highly selective use of disruption, it could be employed inappropriately against resilience objectives, perhaps as a low profile compensation for failures to achieve adequate hygiene within public or private large socio-technical systems. Its use could also creep upwards and be less carefully employed if it is used to meet greater volumes of attacks because large private enterprises are loathe to cooperate in hygiene or share critical information about their own vulnerabilities in coordinated cyber defense. If private enterprises do not adequately support their own cyber resilience, they leave the government, the nation, and the DOD to shoulder an unnecessarily heavy burden. In any of these circumstances, the DSOC needs to address the potential role, limitations, and benefits of disruption as a backup to resilience and the potential changes in the current guidance.

Table 2. DSOC Correspondence with Cybered Resilience-Action Requirements (CRR)

1. Redundancy (of knowledge)

CRR 1, "Beyond these recruiting, education, and training initiatives, adoption and scaling of crossgenerational mentoring programs will allow DOD to grow a gifted cyber talent base for future defense and national security missions.", Strategic Initiative 5, p.11

CRR 1, "The development and retention of an exceptional cyber workforce is central to DOD's success in cyberspace and each of the strategic initiatives outlined in this strategy.", Strategic Initiative 5, p.10

2. Slack (in time to respond)

CRR 2, "[T]o deter and mitigate insider threats, DOD will strengthen its workforce communications, workforce accountability, internal monitoring, and information management capabilities.", Strategic Initiative 2, p.6

3. Organizational Discovery Trial-and-Error Learning (TEL)

CRR 3, " A cornerstone of this activity will be the inclusion of cyber red teams throughout war games and exercises.", Strategic Initiative 1, p.6

CRR 3, "Manage cyberspace risk through efforts such as increased training, information assurance, greater situational awareness, and creating secure and resilient network environments", Strategic Initiative 1, p.5

4. Collective Sense-making,

CRR 4 , "DOD's information technology needs—from modernizing nuclear command and control systems to updating word-processing software—will adopt differing levels of oversight based on the Department's prioritization of critical systems", Strategic Initiative 5, p.11

CRR 4, "Assure integrity and availability by engaging in smart partnerships, building collective self defenses, and maintaining a common operating picture", Strategic Initiative 1, p.5

CRR 4, "DOD is also establishing a pilot public-private sector partnership intended to demonstrate the feasibility and benefits of voluntarily opting into increased sharing of information about malicious or unauthorized cyber activity and protective cybersecurity measures.", Strategic Initiative 3, p.8

CRR 4, "Paradigm-shifting approaches such as the development of Reserve and National Guard cyber capabilities can build greater capacity, expertise, and flexibility across DOD, federal, state, and private sector activities.", Strategic Initiative 5, p.11

5. Rapid Accurate Mitigation, Improvisation, and Adaptation Action,

CRR 5, "DOD will be willing to sacrifice or defer some customization to achieve speedy incremental improvements", Strategic Initiative 5, p.11

CRR 5, "DOD will catalyze U.S. scientific, academic, and economic resources to build a pool of talented civilian and military personnel to operate in cyberspace and achieve DOD objectives.", Strategic Initiative 5, p.10

CRR 5, "DOD's acquisition processes and regulations must match the technology development life cycle. With information technology, this means cycles of 12 to 36 months, not seven or eight years....DOD will employ incremental development and testing rather than a single deployment of large, complex systems.", Strategic Initiative 5, p.11

CRR 5, "Ensure the development of integrated capabilities by working closely with Combatant Commands, Services, Agencies, and the acquisition community to rapidly deliver and deploy innovative capabilities where they are needed the most.", Strategic Initiative 1, p.5

6. Frequent Whole System Practice for Extreme Events under Urgent Conditions

CRR 6, " DOD has had limited capability to simulate cyberspace operations. The National Cyber Range, which allows the rapid creation of numerous models of networks, is intended to enable the military and others to address this need by simulating and testing new technologies and capabilities.", Strategic Initiative 5, p.12

7. Enforceable System Cyber Hygiene (to include integrity-assured component suppliers),

CRR 7, "Additionally, increases in the number of counterfeit products and components demand procedures to both reduce risk and increase quality.", Strategic Initiative 3, p.9

CRR 7, "DOD is enhancing its cyber hygiene best practices to improve its cybersecurity.", Strategic Initiative 2, p.6

CRR 7, "DOD's efforts will focus on communication, personnel training, and new technologies and processes. DOD seeks to foster a stronger culture of information assurance within its workforce to assure individual responsibility and deter malicious insiders by shaping behaviors and attitudes through the imposition of higher costs for malicious activity.", Strategic Initiative 2, p.7

CRR 7, "Improved security measures will be taken with all of the systems that DOD buys, including software and hardware. No backdoor can be left open to infiltration; no test module can be left active. These principles will be a part of, and reinforced by, DOD's trusted defense systems and supply chain risk mitigation strategies. For its hardware, software, architecture, systems, and processes, DOD will take a security in depth approach to design, acquisition, and implementation of trustworthy systems.", Strategic Initiative 5, p.11

8. Underlying Technology-Secure Design Transformation

CRR 8, "Fourth, DOD is developing new defense operating concepts and computing architectures.", Strategic Initiative 2, p.6

CRR 8, *"Technological innovation is at the forefront of national security, and DOD will foster rapid innovation and enhance its acquisition processes to ensure effective cyberspace operations. DOD will invest in its people, technology, and research and development to create and sustain the cyberspace capabilities that are vital to national security.",* Strategic Initiative 5, p.10

9. Comprehensive Multi-Organizational/Layer Learning for Systemic Generative Innovation,

CRR 9, *"An enhanced partnership between DHS and DOD will ... improve a shared understanding of cybersecurity needs and ensure the protection of privacy and civil liberties. Third, the arrangement will conserve limited budgetary resources.",* Strategic Initiative 3, p.8

CRR 9, *"To encourage private sector participation in the development of robust cyberspace capabilities, DOD will empower organizations to serve as clearing houses for innovative concepts and technologies, rewarding firms that develop impactful and innovative technologies.",* Strategic Initiative 5, p.12

10. Stratified Two-Way Flow Sensors/Tagged-linked Interoperable Policy-guided Gateways ("National Cyber Borders")

CRR 10, *"Third, DOD will employ an active cyber defense capability to prevent intrusions onto DOD networks and systems.",* Strategic Initiative 2, p.6

11. Extensive Wicked Actor(s) OODA Loop/Business Model/Motivation Knowledge Collection and Development,

CRR 11, *"DOD is also partnering with the Defense Industrial Base (DIB) to increase the protection of sensitive information. .... To increase protection of DIB networks, DOD launched the Defense Industrial Base Cyber Security and Information Assurance (CS/IA) program in 2007.",* Strategic Initiative 3, p.8

CRR 11, *"DOD will continue to support the development of whole-of-government approaches for managing risks associated with the globalization of the information and communications technology sector. Many U.S. technology firms outsource software and hardware factors of production, and in some cases their knowledge base, to firms overseas.",* Strategic Initiative 3, p.9

CRR 11, *"The development of international shared situational awareness and warning capabilities will enable collective self-defense and collective deterrence. By sharing timely indicators about cyber events, threat signatures of malicious code, and information about emerging actors and threats, allies and international partners can increase collective cyber defense.",* Strategic Initiative 4, p.9

12. Selected Controlled Disruption under Protective Principle of International Law

CRR 12, *"[DOD] reserve[s] the right to defend these vital national assets as necessary and appropriate. These efforts will sustain a cyberspace that provides opportunities to innovate and yield benefits for all", Strategic Initiative 4, p.10*

13. Collective Understandings/Undertakings with Like-minded Cyber Responsible States

CRR 13, *"Engagement will create opportunities to initiate dialogues for sharing best practices in areas such as forensics, capability development, exercise participation, and public-private partnerships.", Strategic Initiative 4, p.10*

CRR 13, *"The Department will work with interagency and international partners to encourage responsible behavior and oppose those who would seek to disrupt networks and systems, dissuade and deter malicious actors.", Strategic Initiative 4, p.10*

Conclusion: National Surprise-Resilience on a Path to a Cyber-Westphalian World

Cyberspace, from its inception, has been about the value, ownership, propagation, and generative[48] capacity of knowledge assets held or developed anywhere that the global substrate reaches. Knowledge is what is stolen, denied, controlled, left undiscovered, or developed into comparative advantage. Knowledge reduces surprises or ensures them when used to disturb processes that are already prone to surprise. At the end of the day, conflict in, through, around, and enabled by cyberspace is a fight over knowledge that is critical to the national existence and economic wellbeing of the defending nations.

Nations without resilience able to protect the knowledge processes and stocks critical to state socio-technical and economic wellbeing will lose cybered conflicts repeatedly. Without resilience in a complexly cybered world, even the soft power steps identified by Nye will be made more difficult internationally. As Nye has noted on several occasions, during the Cold War, the U.S. and the Soviets had to learn together what was needed to be mutually safe.[49] Learning how to be resilient and staying that way are precisely what the counter-resilience strategy of wicked actors would like to inhibit. It is all too easy in the stealthy, noisy, mass volume world of cyberspace to miss the initiating and continuing signals of cybered conflict. If democratic nations have not reoriented their national security focus away from the legacy fixations on offense, defense, and kinetic signs of aggression, their leaders may not even realize they have been in a cybered conflict with multiple actors successfully executing counter-resilience operations for a number of years.

In cyberspace, it is possible to lose the knowledge assets by which you can know you are losing in a cybered conflict until it suits the adversaries to make that point overtly. In short, a state could be so knowledge-poor that it does not have the information or resources necessary to negotiate, even in small steps, toward mutual learning with other nations or major actors in cyberspace. The strategic focus of any democratic state hoping to survive well given the complexity across the world's cyberspace substrate should be developing and acting on the knowledge needed to accommodate the range of surprise generators from normal accidents to wicked actors. "Resilience always, disruption when necessary" should be an underlying motto of all strategic options and seen as a crucial national security priority.

Chris Demchak is Professor in the Strategic Research Department of the U.S. Naval War College and co-director of NWC's Center for cyber conflict studies, and China cyber conflict group. An early member of the Intelligence and Security Informatics (ISI) research field, Dr. Demchak has taught undergraduate and graduate level courses on comparative security and modernized organizations, the institutional history of war and the state, the emerging global information systems, and the worldwide diffusion of defense technologies to include the use of game-based simulations in security analysis. She is currently working on a book manuscript entitled tentatively *Cybercommands: National Responses to Uncertainty and Cybered Conflict* and occasionally contributes to the ACUS "Cybered Conflict" blog. Her research focus is the evolution in organizations, tools, social integrations, and range of choices emerging in Westernized nations' cybersecurity/deterrence strategies, creations or adaptations of cybercommands or equivalents, and institutionalized organizational learning after experiences with cybered conflict and cascading surprise. She has published numerous articles on societal security difficulties with largescale information systems to include cyberwar and cyber privacy, security institutions, and new military models, as well as several books: *Military Organizations, Complex Machines* in the Cornell Security Studies series, *Designing Resilience* (co-edited), and recently, *Wars of Disruption and Resilience: Cybered Conflict, Power and National Security* (UGA Press International relations and technology series). Dr. Demchak received a Ph.D. from Berkeley in political science with a focus on organization theory, security, and surprise in complex socio-technical systems across nations. She also holds two masters degrees, respectively, in economic development (Princeton) and energy engineering (Berkeley).

Appendix A: Mallery 'Thought Experiment' on Securing a Border in Cyberspace

" [D]igital borders in cyber space address remote access over networks [and not close access or supply chain attacks]. ...If bits were tagged to any degree they would reduce the difficulty of the defender in traceback, present a [higher "work factors"[50]] hurdle to attackers, and strengthen incentives for better network and system hygiene. ... [T] tagging operations ... [can] be performed using labeling techniques that cryptographically certify the source and the integrity of the payload, whether a discrete or streamed resource. The same mechanism is applied at the level or the Internet Service Provider (ISP) within countries. And, when enterprises choose to participate, the same mechanisms are applied at their level........ Now, it becomes possible to check at boundaries of countries, IPSs and participating organizations where traffic came from in general terms and whether the content had been altered in transit.

... Transit traffic assertions can be checked against the exit certification of the sender and the entry certification of the receiver, and therefore, this scheme cannot be spoofed without collusion with the sender or receiver. .... Digital borders put in place interfaces that allow states, ISPs and participating enterprises to implement borders at Internet data rates because routers can be designed to check tags very fast, much faster than deep packet inspection (DIP). The administrative entity in control of the border router decides what traffic they accept, forward or send under what circumstances ... But most importantly, it makes the large entities in cyber space accountable for the traffic that they admit and emit.[and facilitates the creation of] .. real-time reputations. Now, users and ISPs may set their origin hygiene preferences to decide under what circumstances they are prepared to receive traffic from zones with high malicious density. Thus, this framework implements a distributed incentive structure that motivates large network actors to improve hygiene within their administrative zones, or at least not emit bad traffic, ... States might distinguish between various kinds of traffic [eg, anonymous, commercial, education , diplomatic, military, etc ...

.... A digital borders architecture like this can be deployed in stages building out from like-minded countries. It will drive development of secure routers and other supporting infrastructure, opening opportunities for innovation in the network equipment sector. It will also drive the development and deployment of host architectures capable of enforcing separation of domains, ... [B]ottom-up or top-down approaches [could] produce the same outcome...[by building] out from fine-grained tagging in hosts designed to enforce separation within enterprise administrative zones. All transit policies would be defined at the enterprise [or ISP] level, which would become the surrogate for national borders in cyberspace.... ...The notions for how digital borders might be implemented are preliminary and require more detail technical analysis to validate feasibility.[51]

Appendix B. Atrium Model of Organizational Learning for Urgent Complex Surprise Resilience (Demchak Proposal)

One approach, the Atrium model, produces collective empowerment against surprise by engaging all members of the defending organization in tacit knowledge collection, hypothesis testing, and easily shared playing through of plausible solution combinations. The basic ATRIUM organization's knowledge base is actively nurtured both in the humans and in the digitized institutional integrated structure because its members 'play through' their security concerns and hypotheses iteratively, not only sharing what is known but collectively uncovering what has not yet been recognized.

One "goes into" the Atrium portion of the resilient organization as a consumer, contributor, or producer – each organization or individual cycles through roles. As each person transfers into a new position, becomes accustomed to the new office and work, operates, and then moves on, each player spends several weeks doing a tacit data dump, including frustrations about process, data, and ideas, into their organization's share of the Atrium files. They then engage in four to five days of playing through their hypotheses in high fidelity, online, avatar-based, continuously available, co-authored game-like simulations using real operational data. The graphical interface alone engenders human abilities to see reality, their reality, and alter it to test ideas, improve prior choices, or learn with others about combinatorial novel responses. The recording ability capturing all choices, conversation, considerations, changes, and replays means others can review the 'game' for lessons learned or, using their own tacit knowledge, replay it to a new outcome. Identifying tags can be flexible and even on occasion masked to encourage honesty and then the knowledge is added to the central pools. System organizational members elsewhere can then apply data mining or other applications on this expanding pool of knowledge elements to guide their future processes.

Explicit and implicit comparative institutional knowledge thus becomes instinctively valued and actively retained and maintained for use in ongoing or future operations. While everyone cycles through the Atrium routinely to download experiences, every so often, perhaps once every six months, each person also spends a week or so setting up questions and looking at the data for all the system's benefit.

National resilience requires this organizational learning ability to routinely play through the what-if questions of organizations and policymakers on a scale and fidelity commensurate with the surprises possible in the cybered system. Their experiments benefit themselves and their organization when they record and replay choices to learn by simulated discovery trial and error.

This model also offers nations a way to develop national resilience across joint or cooperating institutions, public or private. Rotating into the Atrium, members of varying enterprises can at any time play through confounding scenarios with players from other organizations while using institutionally sanitized but current operational data. The collective learning experience extends beyond the first round of players to all others using the replays or results in data mining, scanning for alternatives under urgent conditions, or doing analysis of consequences individually or collectively. In the process intra and inter- organizational trust is enhanced, as well as the spread of a wider view of other entities capacities and constraints.

Bibliography

Allen, P. D., and C. C. Demchak. "The Palestinian-Israeli Cyberwar." *Military Review* 83, no. 2 (2003): 52-59.

Arquilla, J. "How to Lose a Cyberwar." *Foreign Policy online*, December 12 2009.

Bar-Yam, Y. *Dynamics of Complex Systems*. Westview Pr, 2003.

Baskerville, R. "Hacker Wars: E-Collaboration by Vandals and Warriors." *International Journal of e-Collaboration* 2, no. 1 (2006): 16.

Benedikt, M. *Cyberspace: First Steps*. MIT Press, 1991.

Buchanan, R. "Wicked Problems in Design Thinking." *Design issues* (1992): 5-21.

Buxbaum, P.A., and MIT Correspondent. "Battling Botnets." *Military Information Technology*, May 13 2010.

Castells, M. *Power of Identity: The Information Age: Economy, Society, and Culture*. Cambridge, MA, USA: Blackwell Publishers, Inc. , 1997.

Comfort, L. A. Boin, and C. C. Demchak, eds. *Designing Resilience: Preparing for Extreme Events*. Pittsburgh: University of Pittsburgh Press, 2010.

De Bruijne, M., and M. Van Eeten. "Systems That Should Have Failed: Critical Infrastructure Protection in an Institutionally Fragmented Environment." *Journal of Contingencies and Crisis Management* 15, no. 1 (2007): 18-29.

Demchak, C. C. ""Atrium"–a Knowledge Model for Modern Security Forces in the Information and Terrorism Age." *Lecture Notes In Computer Science* (2003): 223-31.

Demchak, C. C. "Complexity, Rogue Outcomes and Weapon Systems." *Public Administration Review* 52, no. 4 (1992): 347-55.

Demchak, C. C. "Lessons from the Military: Surprise, Resilience, and the Atrium Model." Chap. 4 In *Designing Resilience: Preparing for Extreme Events*, edited by L. Comfort, A. Boin and C. Demchak. 62-83. Pittsburgh: University of Pittsburgh Press, 2010.

Demchak, C. C. *Wars of Disruption and Resilience: Cybered Conflict, Power, and National Security*. Athens, Georgia, USA: University of Georgia Press, 2011.

Demchak, C. C., and P. J. Dombrowski. "Rise of a Cybered Westphalian Age". *Strategic Studies Quarterly* 5, no. 1 (March 1 2011): 31-62.

Economist Staff. "The Future of the Internet: A Virtual Counter-Revolution." *The Economist*, September 2 2010.

Erickson, J. *Hacking: The Art of Exploitation*. New York: No Starch Press, 2008.

Falliere, N., L. O. Murchu, and E. Chien. "W32.Stuxnet Dossier: Version 1.3." online : http://www.symantec.com/content/en/us/enterprise/media/security_response/whitepapers/w32_stuxnet_dossier.pdf: Symantex Inc, 2010.

Galbraith, J. "Organizational Design." *Reading, MA* (1977).

Gleick, J. *Chaos: The Making of a Science*. New York: Vintage Books, 1988.

Goldman, D. "The Cost of Cybercrime -- the Price Tag on Corporate Data Breaches Is Soaring: The Rise in Cybercrime Is Costing Hundreds of Billions of Dollars Each Year." *CNNMoney.com*, July 22 2011.

Gomory, R. E. "An Essay on the Known, the Unknown and the Unknowable." *Scientific American* 272 (1995): 120.

Gross, M. J. "Stuxnet Worm: A Declaration of Cyber-War." *Vanity Fair*, April 2011, online.

Heimann, C.F.L. "Understanding the Challenger Disaster: Organizational Structure and the Design of Reliable Systems." *The American Political Science Review* 87, no. 2 (1993): 421-35.

Hille, K. "How China Polices the Internet." *Financial Times* online, July 17 2009.

Jervis, R. *System Effects: Complexity in Political and Social Life*. Princeton University Press, 1997.

Kennedy, P.M. *The Rise and Fall of the Great Powers*. Random House New York, 1988.

Kumar, R., P. Raghavan, S. Rajagopalan, and A. Tomkins. "Trawling the Web for Emerging Cyber-Communities." *Computer networks* 31, no. 11-16 (1999): 1481-93.

LaPorte, T.R. *Organized Social Complexity: Challenge to Politics and Policy*. Princeton University Press, 1975.

Leveson, N., E. Hollnagel, and D. D. Woods. *Resilience Engineering: Concepts and Precepts*. Ashgate Publishing, Ltd., 2006.

Libicki, M.C. *Cyberdeterrence and Cyberwar*. Washington DC: Rand Corporation, 2009.

Mallery, J. C. "A Strategy for Cyber Defense (Earlier Title: Multi-Spectrum Evaluation Frameworks and Metrics for Cyber Security and Information Assurance)." In *MIT/Harvard Cyber Policy Seminar,*. Cambridge, MA: Massachusetts Institute of Technology Computer Science & Artificial Intelligence Laboratory 2011 (2009).

Markoff, J., D. E. Sanger, and T. Shanker. "Cyberwar: In Digital Combat, U.S. Finds No Easy Deterrent." *New York Times*, January 26 2010.

McDaniel, R. R., and D. J. Driebe. *Uncertainty and Surprise in Complex Systems: Questions on Working with the Unexpected*. Springer, 2005.

Montrose, L. *War through the Ages*. New York: Harper & Bros, 1944.

Murphy, M. "Cyberwar:War in the Fifth Domain." *The Economist online*, July 1 2010.

Norton, L. P. "Dragon Fire." *Barron's*, June 27 2011.

Nye Jr, J.S. *The Future of Power in the 21st Century*. Cambridge, MA: Public Affairs, 2011.

Patel, N. "Breaking News: Botched Mcafee Update Shutting Down Corporate Xp Machines Worldwide." *Engadget Online*, April 21 2011.

Perrow, C. *Normal Accidents*. Basic Books New York, 1984.

Randel, J. M., B. A. Morris, C. D. Wetzel, and B. V. Whitehill. "The Effectiveness of Games for Educational Purposes - a Review of Recent Research." [In English]. *Simulation & Gaming* 23, no. 3 (Sep 1992): 261-76.

Rheingold, H. "Virtual Communities: Homesteading on the Electronic Frontier." *Reading: Addison Wesley* (1993).

Rochlin, G. I. "Reliable Organizations: Present Research and Future Directions." *Journal of Contingencies and Crisis Management* 4, no. 2 (1996): 55-59.

Russell, F. S. *Information Gathering in Classical Greece*. Ann Arbor: University of Michigan Press, 1999.

Segal, A. "Pentagon's New Strategy for Operating in Cyberspace Breaks Little Ground & Offers Few Specifics." *Council on Foreign Relations*, Twitter Feed, July 21 2011.

Shiode, N. "Toward the Construction of Cyber Cities with the Application of Unique Characteristics of Cyberspace." *Online Planning Journal* (1997).

Shooman, M. L. *Reliability of Computer Systems and Networks*. London: Wiley-Interscience [Great Britain], 2002.

Smith, M.A., and P. Kollock. *Communities in Cyberspace*. Psychology Press, 1999.

Sunstein, C. R. "Irreversible and Catastrophic." *Cornell Law Review* 91, no. 4 (May 2006): 841-97.

Van Bueren, E.M., E.H. Klijn, and J.F.M. Koppenjan. "Dealing with Wicked Problems in Networks: Analyzing an Environmental Debate from a Network Perspective." *Journal of Public Administration Research and Theory* 13, no. 2 (2003): 193.

Waltz, K. N. *Theory of International Politics*. McGraw-Hill, 1979.

Wohl, J.G. "Force Management Decision Requirements for Air Force Tactical Command and Control." *IEEE TRANS. SYS., MAN, AND CYBER*. 11, no. 9 (1981): 618-39.

Youngblood, P., P. Harter, S. Srivastava, W. LeRoy Heinrichs, and P. Dev. "A Virtual Learning Environment for Team Training in Trauma Management." *Acad Emerg Med* 12, no. 8 (August 1, 2005 2005): 792-b-93.

Zittrain, J.L. "The Generative Internet." *Harvard Law Review* (2006): 1974-2040.

[1] I would like to express my appreciation to John Mallery for his comments on this work. All errors are mine, of course.

[2] Benedikt, M. Cyberspace: First Steps. MIT Press, 1991.

[3] Shiode, N. "Toward the Construction of Cyber Cities with the Application of Unique Characteristics of Cyberspace." *Online Planning Journal* (1997). Castells, M. *Power of Identity: The Information Age: Economy, Society, and Culture*. Cambridge, MA, USA: Blackwell Publishers, Inc. , 1997.

[4] Kumar, R. P. Raghavan, S. Rajagopalan, and A. Tomkins. "Trawling the Web for Emerging Cyber-Communities." *Computer networks* 31, no. 11-16 (1999): 1481-93. Smith, M.A., and P. Kollock. *Communities in Cyberspace*. Psychology Press, 1999.

[5] Waltz, K. N. *Theory of International Politics*. McGraw-Hill, 1979. Montrose, L. *War through the Ages*. New York: Harper & Bros, 1944.

[6] In this piece, 'knowledge assets' are a shorthand based on the socio-technical systems approach to organizational surprise and complexity. Knowledge in this use includes such assets as capital; information of any sort that is removable, to include plans, passwords, or raw data; remote controls of missile systems or stock markets as a knowledge; foreknowledge of a sabotaged system or information about corrupted supply components, remotely or insider inserted programs successfully installed and

confirmation conveyed to attackers; or forms of social capital in ideas or innovations not yet in the public view. Knowledge constitutes the certainty that transfers to attackers when knowledge assets are taken. (Demchak 1991)

[7] Kennedy, P.M. *The Rise and Fall of the Great Powers.* Random House New York, 1988.

[8] I discuss the search for knowledge in order to disrupt enemy city-states, specifically Sparta and Athens, in a recent book. See chapter 2 in Demchak, C. C. *Wars of Disruption and Resilience: Cybered Conflict, Power, and National Security.* Athens, Georgia, USA: University of Georgia Press, 2011.

[9] Available at *www.cio.bund.de/SharedDocs/.../DE/.../css_engl_download.pdf.*

[10] A reference to a critical major last stand of the ancient Greeks against an overwhelming Persian invading army. The French cybersecurity strategy is available at *http://www.enisa.europa.eu/media/news-items/french-cyber-security-strategy-2011.*

[11] Available at *www.direct.gov.uk/prod_consum_dg/groups/.../dg_191639.pdf.* Note also that the UK issued a new cybersecurity strategy in fall 2011 to reinforce the importance of cyberspace as a top tier national security concern.

[12] Internet Protocol version 6 has exponentially more addresses than the current standard, IPv4, which is limited to approximately 4.3 billion. The larger number will profoundly affect connecting and tracking all cybered devices worldwide.

[13] Private observations from cyber, defense, and business analysts that without the so-called "cyber Pearl Harbor," nations will not move as they must to change the underlying insecurity of this substrate. As recently as May 2011, a cybersecurity expert commented in a private conversation with the author that his job to protect his corporation is constantly difficult because "they [leaders of the organization] have to be personally hurt really hard for them to care".

[14] Falliere, N., L. O. Murchu, and E. Chien. "W32.Stuxnet Dossier: Version 1.3." online : http://www.symantec.com/content/en/us/enterprise/media/security_response/whitepapers/w32_stuxnet_dossier.pdf: Symantex Inc, 2010.

[15] Gross, M. J. "Stuxnet Worm: A Declaration of Cyber-War." *Vanity Fair*, April 2011, online.

[16] Russell, F. S. *Information Gathering in Classical Greece.* Ann Arbor: University of Michigan Press, 1999.

[17] Demchak, C. C. "Complexity, Rogue Outcomes and Weapon Systems." *Public Administration Review* 52, no. 4 (1992): 347-55.

[18] LaPorte, T. R. *Organized Social Complexity: Challenge to Politics and Policy.* Princeton University Press, 1975; Rochlin, G. I. "Reliable Organizations: Present Research and Future Directions." *Journal of Contingencies and Crisis Management* 4, no. 2 (1996): 55-59. Complexity matters greatly the more the complex system is critical to survival and cannot afford to be surprised in catastrophically nasty ways. Sunstein, C. R. "Irreversible and Catastrophic." *Cornell Law Review* 91, no. 4 (May 2006): 841-97. Without any mistakes or maliciousness by humans, large complex systems in critical roles will have a routine level of "normal accidents" that is tied to the level of complexity of the overarching system. Perrow, C. *Normal Accidents.* Basic Books New York, 1984. Surprises in large complex systems of standardized components can in some form or frequency be foreseeable, and whole swathes of the engineering community dedicate their research to devising responses to sudden and surprising precision failures and cascading disruptions inside large-scale technical systems. Leveson, N., E. Hollnagel, and D. D. Woods. *Resilience Engineering: Concepts*

and Precepts. Ashgate Publishing, Ltd., 2006. While it is often enough true that some surprises are due to neglect because the foreknowledge and accommodation efforts are considered too hard and expensive in large complex systems, failing capacitors do not change their activities when they are observed to be failing. In broad engineering terms, some systems designs can make these surprises less likely. Shooman, M. L. *Reliability of Computer Systems and Networks*. London: Wiley-Interscience [Great Britain], 2002. That is, the systems dealing with normal accidents, à la Perrow, can be made more knowable. Surprise accommodation choices for such systems begin with the initial designs that incorporate redundancy of knowledge (not necessarily specific components), slack in the calibration or timing of responses in order to accommodate imprecise inputs, and extensive stress-compensate-retest iterations in advance and during the lifespan of the technological system. Heimann, CFL. "Understanding the Challenger Disaster: Organizational Structure and the Design of Reliable Systems." *The American Political Science Review* 87, no. 2 (1993): 421-35. These choices help but do not eliminate the rogue outcomes, those surprises that are both unknowable in advance and not serendipitously accommodated by choices not directly related to the surprise. Gomory, R. E. "An Essay on the Known, the Unknown and the Unknowable." *Scientific American* 272 (1995): 120. But they can accommodate and reduce the harm of more knowable failures and their cascades. Resilience engineering translates these lessons of redundancy into more reliable engineering systems design and operation. Leveson, Hollnagel, and Woods, *Resilience Engineering: Concepts and Precepts*.

[19] Patel, N. "Breaking News: Botched Mcafee Update Shutting Down Corporate Xp Machines Worldwide." *Engadget Online*, April 21 2011.

[20] McDaniel, R. R., and D. J. Driebe. *Uncertainty and Surprise in Complex Systems: Questions on Working with the Unexpected*. Springer, 2005.

[21] For systems incorporating humans directly into their operations as large socio-technical systems (LTS) that cannot fail or must at least gracefully degrade, there is a similar, though smaller, literature on accommodating surprises. Humans and complex technological systems are merged in critical societal functions. This literature goes beyond initial organizational designs to organizational learning methods enacted in advance of crisis conditions, requiring knowledge development in thoughtful institutional design and senior consistent sustaining management support. De Bruijne, M., and M. Van Eeten. "Systems That Should Have Failed: Critical Infrastructure Protection in an Institutionally Fragmented Environment." *Journal of Contingencies and Crisis Management* 15, no. 1 (2007): 18-29. The organizational designs (including processes, buy-in, authorities, etc) must engender, sustain, and innovate through the collective efforts by the whole of the affected community to have the knowledge, redundancy, and slack capacities where, when, and at the effectiveness needed when surprised. Comfort, L., A. Boin, and C. C. Demchak, eds. *Designing Resilience: Preparing for Extreme Events*. Pittsburgh: University of Pittsburgh Press, 2010.

[22] Author's private conversation with former Estonian Ministry of Defense official, summer 2011.

[23] Goldman, D. "The Cost of Cybercrime--the Price Tag on Corporate Data Breaches Is Soaring: The Rise in Cybercrime Is Costing Hundreds of Billions of Dollars Each Year." *CNNMoney.com*, July 22 2011.

[24] Galbraith, J. "Organizational Design." *Reading, MA* (1977).

[25] In complex systems, these kinds of unforeseeable problems are estimated by Wohl to be between 5 and 20 percent of the outcomes, depending on the complexity of the system and what knowable information was left unsought by the designers and operators of a complex system. Wohl, JG. "Force Management Decision Requirements for Air Force Tactical Command and Control." *IEEE TRANS. SYS., MAN, AND CYBER.* 11, no. 9 (1981): 618-39. See also the seminal book by Gleick on complexity and the more recent

works by Bar-Yam and McDaniel. Gleick, J. *Chaos: The Making of a Science*. New York: Vintage Books, 1988. McDaniel and Driebe, *Uncertainty and Surprise in Complex Systems: Questions on Working with the Unexpected*. Bar-Yam, Y. *Dynamics of Complex Systems*. Westview Pr, 2003.

26 Van Bueren, E.M., E.H. Klijn, and J.F.M. Koppenjan. "Dealing with Wicked Problems in Networks: Analyzing an Environmental Debate from a Network Perspective." *Journal of Public Administration Research and Theory* 13, no. 2 (2003): 193. Buchanan, R. "Wicked Problems in Design Thinking." *Design issues* (1992): 5-21.

27 Observation, Orientation, Decision, Action; a reference to the cycle of recognition and action needed by military organizations in responding to actions by enemies. First developed by John Boyd in the 1980s for the U.S. Air Force and widely used in the U.S. military to indicate the need to react more quickly and use feedback more accurately than an enemy is able to.

28 These three map across a number of social science literatures, from international relations (God, butter, guns; or constructivism, liberal institutionalism, realism) to gang warfare (referent group, life options, ego) to social relations (religion, wealth, empowerment). See forthcoming book. Demchak, C. C. *Wars of Disruption and Resilience: Cybered Conflict, Power, and National Security.*

29 Galbraith, J. R. *Organization Design*. Addison-Wesley Reading, Mass, 1977.

30 Ibid.

31 Economist Staff. "The Future of the Internet: A Virtual Counter-Revolution." *The Economist*, September 2, 2010.

32 Demchak, C. C., and P. J. Dombrowski. "Rise of a Cybered Westphalian Age". *Strategic Studies Quarterly* 5, no. 1 (March 1 2011): 31-62.

33 Nye Jr, J.S. *The Future of Power in the 21st Century*. Cambridge, MA: Public Affairs, 2011.

34 Mallery, J. C. "Traffic Tainting at Boundaries: A Thought Experiment in Cyber Defense," Cambridge: MIT Computer Science & Artificial Intelligence Laboratory, draft technical memo, July 25, 2011.

35 Hille, K. "How China Polices the Internet." *Financial Times* online, July 17 2009.

36 See for example, Tikk, E. 2011. "10 Rules of Behavior for Cyber Security." www.CCDCOE.org

37 Baskerville, R. "Hacker Wars: E-Collaboration by Vandals and Warriors." *International Journal of e-Collaboration* 2, no. 1 (2006): 16.

38 I am grateful to John Mallery for pointing out that one could conceive of broad spectrum distributed network disruption against bad actors and crowd disruption. However, for the purposes of controlling the downstream and collateral effects, it is not clear these choices ease the knowledge requirements that are so burdensome.

39 See Libicki on the characteristics of a cyberwar. Libicki, M.C. *Cyberdeterrence and Cyberwar*. Washington DC: Rand Corporation, 2009.

40 Buxbaum, P.A., and MIT Correspondent. "Battling Botnets." *Military Information Technology*, May 13 2010.

41 This equation comes from the 'theory of action' argument concerning disruption made in "Wars of Disruption and Resilience" book by author. Demchak, *Wars of Disruption and Resilience: Cybered Conflict, Power, and National Security.*

[42] Allen, P. D., and C. C. Demchak. "The Palestinian-Israeli Cyberwar." *Military Review* 83, no. 2 (2003): 52-59.

[43] Author's personal communication with John Mallery, MIT. See also Mallery, J. C. "A Strategy for Cyber Defense (Earlier Title: Multi-Spectrum Evaluation Frameworks and Metrics for Cyber Security and Information Assurance)." In *MIT/Harvard Cyber Policy Seminar,*. Cambridge, MA: Massachusetts Institute of Technology Computer Science & Artificial Intelligence Laboratory 2011 (2009).

[44] Arquilla, J. "How to Lose a Cyberwar." *Foreign Policy* online, December 12 2009. Murphy, M. "Cyberwar: War in the Fifth Domain." *The Economist* online, July 1 2010; Jervis, R. *System Effects: Complexity in Political and Social Life.* Princeton University Press, 1997.

[45] To be fair, the international hacking community now routinely adds "a good discovery process" to their definition of the social positive aspect of hacking, that it should be done to help the owners of the network, presumably as requested. Erickson, J. *Hacking: The Art of Exploitation.* New York: No Starch Press, 2008.

[46] Gross, M. J. "Stuxnet Worm: A Declaration of Cyber-War."

[47] Segal, A. "Pentagon's New Strategy for Operating in Cyberspace Breaks Little Ground & Offers Few Specifics." *Council on Foreign Relations,* Twitter Feed, July 21 2011.

[48] Generativity is a term developed by Zittrain, indicating the ability of networked systems to develop new knowledge. Zittrain, J.L. "The Generative Internet." *Harvard Law Review* (2006): 1974-2040.

[49] Nye Jr, J. S. *The Future of Power in the 21st Century.*

[50] See Mallery, J. C. "A Strategy for Cyber Defense (Earlier Title: Multi-Spectrum Evaluation Frameworks and Metrics for Cyber Security and Information Assurance)."

[51] Mallery, J. C., "Draft Technical Memo: Traffic tainting at boundaries: a thought experiment in cyber defense." Cambridge: MIT, 2011.

[52] Material drawn and revised from previously published material on the Atrium model. For the most recent discussion, see Demchak, C. C. "Lessons from the Military: Surprise, Resilience, and the Atrium Model." Chap. 4 In *Designing Resilience: Preparing for Extreme Events,* edited by Comfort, L., A. Boin and C. C. Demchak. 62-83. Pittsburgh: University of Pittsburgh Press, 2010.

[53] Allen and Demchak, "The Palestinian-Israeli Cyberwar."

[54] Demchak, C. C. "Atrium"–a Knowledge Model.for Modern Security Forces in the Information and Terrorism Age." *Lecture Notes In Computer Science* (2003): 223-31.

[55] Randel, J. M., B. A. Morris, C. D. Wetzel, and B. V. Whitehill. "The Effectiveness of Games for Educational Purposes - a Review of Recent Research." [In English]. *Simulation & Gaming 23,* no. 3 (Sep 1992): 261-76.

[56] Youngblood, P., P. Harter, S. Srivastava, W. LeRoy Heinrichs, and P. Dev. "A Virtual Learning Environment for Team Training in Trauma Management." *Acad Emerg Med* 12, no. 8 (August 1, 2005 2005): 792-b-93.

Part **2**

CHAPTER 3

Eight Questions and Answers on U.S. Cyber Statecraft

Jason Healey
Director, Cyber Statecraft Initiative
Atlantic Council

CHAPTER 4

Harnessing Leviathan:
Internet Governance and Cybersecurity

James A. Lewis
Senior Fellow and Director
Technology and Public Policy Program
Center for Strategic and International Studies

"With the new International Strategy for Cyberspace, the United States has leapt ahead of other governments, by putting forward a complete vision of cyber statecraft, combining security, intellectual property, Internet freedom, and deterrence. We should expect all future national cyber strategies to become similarly encompassing."

—JASON HEALEY

Eight Questions and Answers on U.S. Cyber Statecraft

Jason Healey
Director, Cyber Statecraft Initiative
Atlantic Council

Over the course of 2011, the United States government released a coordinated set of policies that represents the most energetic cyber statecraft in nearly a decade. This chapter will discuss these policies, their background, implications, and future through a series of eight questions and answers:

1. Is this "cyber" stuff all new? If not, what has changed?

2. What are the outlines of the current administration's cyber policies?

3. What are the current documents of U.S. cyber policy?

4. Who is taking the lead to develop a cohesive framework that will apply across the U.S. government?

5. What are some of the ongoing projects and programs on cyberspace security?

6. What are possible domestic partnerships that can strengthen national capacities?

7. How does this compare to how other governments are organized to approach the problem?

8. What to watch next?

This chapter takes a relatively optimistic outlook on the current answers to these questions, which together give a broad overview of today's U.S. cyber policies and programs. This optimism is unfortunately rooted not in confidence that the projects and initiatives make us more secure, but rather in the progress and fresh thinking that is taking place. We may not be through the policy desert yet, and we may not have reached an oasis, but we have found a glass with some water in it. It may not be full, but we can be optimistic for what lies ahead.

Is this "cyber" stuff all new? If not, what has changed?

Despite the recent headlines announcing the advent of "cyberwar," the problems of computer security and "cyber" are not new.[1] Over the past several decades, numerous reports have indicated that cyberspace is important, even critical, and extremely vulnerable in the face of growing threats from state and non-state actors. Of course, all have called for immediate action. These warnings go as far back as reports from the Defense Science Board in 1970 and the National Research Council in 1991. Additional commissions and boards followed, including Defense Science Board reports in 1996 and 2001, the Marsh Commission (President's Commission on Critical Infrastructure Protection) in 1997, and more recently the Commission on Cybersecurity for the 44th Presidency in 2008.

More striking than the words in these reports are the catastrophes taking place in the networks. Of the thousands of cyber incidents since the late 1980s, at least six were serious enough to be considered "wake up calls."[2] Each had similar underlying causes (vulnerable systems and distracted people) and seized the attention of senior government officials who rightly decided "never again." Yet despite some progress over these intervening decades, the problems highlighted in the reports—subsequently made real in the incidents—remain unsolved: The wake-up calls have been replaced by a snooze bar.

Fortunately, the United States government is in the midst of the third major phase of new policymaking. It is still too early to know if this current policy rollout will lead to any more substantive changes than the last two, in 1997-1998 and 2003.[3] However, there is, as the remaining sections of this chapter will discuss, room for optimism.

What are the outlines of the current administration's cyber policies?

The bad news is that the United States does not have an existing overall cyber strategy. The comprehensive 2003 Strategy to Secure Cyberspace is largely ignored;[4] the 2008 Comprehensive National Cybersecurity Initiative was not comprehensive, focusing solely on government networks; and while the 2009 Cyberspace Policy Review listed ten specific near-term actions, it was not a strategy.[5]

The good news is that there has been far more momentum for strong cyber policies in the last three months than in the last eight years. The outlines of these new policies include:

1. Basic Continuity: These policies generally continue to prioritize cyber efforts in similar ways to past policies. These include strengthening cybersecurity for federal systems, improving protections for consumers, and broadly increasing international cooperation.

2. Some New Ideas: However, this essential continuity should be considered updated, as the U.S. government has learned from the lessons of the past two decades. For example, the Department of Defense is no longer emphasizing offense or deterrence by punishment, and "regulation" is no longer quite as dirty a word as it used to be. Security is increasingly seen as a non-absolute, and indeed the White House has said it is looking for cyberspace to be "secure *enough*" and "reliable *enough*" for it to earn people's trust and "support their work."[6]

3. A Light but Expanding Government Touch: Programs remain generally voluntary, though there is proposed new regulation for companies in critical infrastructure sectors. Nevertheless, the proposed government role is far weaker than called for by some commentators (such as former White House official Richard Clarke or the Commission on Cybersecurity for the 44th Presidency) and would not require Internet Service Providers to provide "clean pipes" relatively free of attacks.

4. Inclusion and Balance of New Areas of Cyber Statecraft: Whereas past cyber strategies typically only covered security and, at times, innovation, the new policies address cybersecurity more holistically. The inclusion of these new areas of cyber statecraft (including norms of international behavior, Internet freedom, and development) allows the government to better prioritize and balance between policies. The DoD has struck a better balance by emphasizing defense over offense, and all cyber strategy documents highlight the importance of the American values of free speech and commerce.

The next section will look at the major documents that give more color to this outline.

What are the current documents of U.S. cyber policy?

The surge of new policies in the last few months includes four highlights: The White House issued its first-ever legislative proposal on cybersecurity along with a strategy to better engage internationally; these documents were herded together through the interagency process along with new cyber strategies from the Departments of Defense and Commerce. This section will first look at the legislative proposal and Commerce strategy (as they are focused on domestic issues), and then at the international and defense strategies (concerned more with issues of national security).

The White House Legislative Proposal[7] has several central elements, including standardizing requirements for reporting to consumers when their personal information may have been compromised; application of racketeering laws to cybercrimes; providing Department of Homeland Security assistance to affected companies and granting immunity to companies sharing information in return; improving transparency for cybersecurity plans for companies in critical infrastructure sectors; and regulating such companies, including mandating outside audits and reporting to the Securities and Exchange Commission. The proposal also updates the main law on the security of federal systems (FISMA); further formalizes and strengthens the role of DHS in federal cybersecurity—including allowing more hiring flexibility and fostering an exchange of experts with the private sector—and makes permanent the DHS authority to oversee intrusion systems across the entire federal government. Lastly, the document proposes a new framework of privacy and civil liberties protection "designed expressly to address the challenges of cybersecurity."[8]

The Department of Commerce's strategy[9] highlights the avoidance of strong regulation, calling for voluntary codes of conduct to decrease vulnerabilities and new incentives to reduce threats, as well as new efforts for consumer education and research for new security technologies. The Department of Commerce is also responsible for implementing an earlier White House strategy for creating "trusted identities" to improve Internet commerce. More recently, the Securities and Exchange Commission has issued guidelines for publicly traded companies to disclose information to investors when they have been subjected to a significant cyber incident.[10]

Moving from domestic affairs to national security, international audiences were often confused by seemingly conflicting statements from the U.S. government that the Internet should be free and yet should be policed for the purposes of security and the protection of intellectual property rights. Moreover, the U.S. military's stated goal of achieving cyber "superiority" or "dominance" served to further obfuscate the U.S. government's views on cybersecurity. The Obama administration's International Cyber Strategy[11] was particularly ground–breaking, as for the first time it combined these uncoordinated and unconnected policies into one, calling for new norms and practices in cybersecurity, while emphasizing traditional American values of free speech, innovation, free trade, and international engagement.

The Department of Defense Strategy for Operating in Cyberspace[12] continues the trend of de-emphasizing the "militarization" of cyberspace, with five initiatives that both normalize and prioritize cyberspace operations. The strategy notes that DoD will treat cyberspace as an operational domain—equal to air, land, sea and space—and calls for new concepts to improve its defenses and embrace "active cyber defenses." It also announces that it will work with partners in the U.S. government, in the private sector, and internationally (starting with traditional allies, such as the "five eyes" partners), and improve the Department's workforce and technology acquisition. The Department of Defense continues to plan to conduct intelligence and offensive operations in cyberspace and still hopes to deter some adversaries by threatening kinetic or cyber retaliation, but these priorities do not feature in the new strategy.

Who is taking the lead to develop a cohesive framework that will apply across the U.S. government?

White House leadership on this issue is far stronger today than it has been since 2003, when much official responsibility passed to the Department of Homeland Security, while institutional power remained with the larger and better-organized Department of Defense (and to a lesser extent, Department of Justice). Since the appointment of Howard Schmidt as the president's cyber coordinator, the National Security Council has been the focus of the U.S. government's efforts; but the Department of State is also newly invigorated, with the appointment of its own cyber coordinator, who reports directly to the secretary. The Department of Defense, meanwhile, has taken a lesser intergovernmental role, especially on international engagement.

Regardless of how good the interagency process is, however, the White House has not issued an updated version of the 2003 cyber strategy to provide overall guidance on cybersecurity. Moreover, the truth is that policy for cyberspace will always lag behind technology. As expressed by Greg Rattray of the Cyber Conflict Studies Association, U.S. government mechanisms resemble those of the industrial age and are poorly equipped to deal with the fast-moving digital economy which spawns disruptive new technologies every few years.[13] This problem is compounded when the digital disruptions are tied to seething national security issues, whether it is unrest on the Arab street, Russian bullying of its near abroad, or aggressive Chinese online espionage.

Barring larger institutional reform, the best the U.S. government interagency process can hope for is to continue to close the gap between "geeks" and "wonks" to help these two tribes intercommunicate.

What are some of the ongoing projects and programs on cyberspace security?

The Department of Homeland Security has put in place many robust initiatives including:

- Strengthening the Department's capability to respond to serious incidents, including hosting representatives of the private sector on its watch and operations floors and engaging senior leadership. At the center of their response efforts are the National Cybersecurity and Communications Integration Center and US CERT, whose role is to coordinate information sharing before and response after significant cyber incidents. DHS has also set up a new response team to focus solely on industrial control systems—the digital devices that control electrical grids, dams, power plants, and factories which are increasingly being connected to the Internet, dramatically increasing the chances of a major cascading failure.

- Pushing intrusion detection (called EINSTEIN 2) and prevention systems (called EINSTEIN 3) throughout the federal government to better detect and stop intrusions. Although DHS has put in place additional initiatives to improve federal cybersecurity—including reducing the number of Internet connections across all government departments and agencies and providing risk management assistance—EINSTEIN is best known, largely because it has been a lightning rod for attention from the media and privacy activists. Whereas EINSTEIN 2 is meant to only detect attacks, which overworked defenders must then respond to, EINSTEIN 3 is designed to more actively and automatically stop attacks while they are underway. In addition, EINSTEIN 3—developed by the National Security Agency—will detect threats using "signatures" based on NSA's classified sources and methods. Accordingly, there are concerns that the system would be used to inappropriately collect information submitted by citizens online to federal agencies (a reasonable concern considering the role of NSA in the wake of the wiretapping scandals). DHS is hoping to begin rolling out the system in 2011 based on a budget request of over $230 million.

Relative Security of .mil, .gov, and .com

There is a sense among policymakers that ".mil is secure, .gov is getting better, and so .com is the problem." That is, the government largely has its act together and therefore must help secure the commercial sector (though whether with carrot or stick remains to be determined).

Unfortunately, this finding would be news to most in the cybersecurity field. First, ".com" is so overly broad as to cover many highly secured parts of the private sector. For example, the leading companies in the finance sector have long paid top dollar for security systems and the top people (often from government) to deploy and operate them. With strong leadership from the Treasury Department and the Federal Reserve Board, the finance sector has in place solid mechanisms for sharing information before attacks and responding together afterwards.

However, this finding is also wrong for a second reason: While there are indeed truly excellent cyber defenses within the U.S. government, notably some parts of the Department of Defense and intelligence community, these are not the norm. Findings from numerous internal and GAO reports find repeated failing of standard, much less best, practice. Departments like Housing and Urban Development (with data on American's home mortgages) and Education (with data on student loans) are too short of resources to adequately protect sensitive information, falling far behind the security measures of the major financial institutions. Even the Department of Defense suffered a major security breach by a private who was able to download data to a compact disc, something forbidden under good security practice. And the DHS Inspector General recently found that US CERT, the center of U.S. cybersecurity, did not conduct the most basic security functions, such as applying security patches, having appropriate documentation, or adhering to department security policies.

With news like this it is often difficult for private sector companies to see the government as fully credible when it comes to cybersecurity. Accordingly, public-private sector partnerships built on "trust us, we're the government" have not often succeeded.

- Better outreach, such as a National Cybersecurity Awareness Month, and a focus on improving the nation's cyber workforce. The DHS also conducts major exercises, with Cyber Storm being the biennial capstone, to test incident response plans and information sharing within government and the private sector.

- Working with DoD, the intelligence community, and others on the Enduring Security Framework dialogue (including classified threat briefings) with major telecommunications providers.

- Working to monitor a limited number of network devices in every one of the fifty states in order to better sample and measure the threats to their networks.

The Department of Defense is pursuing several other major initiatives including:

- The DIB Cyber Pilot, a test program lasting several months, which shares threat information with the defense industrial base (or their service providers). This program does not involve the government "monitoring, intercepting, or storing any private sector communications," according to then-Deputy Secretary of Defense Bill Lynn.[14]

- Active cyber defenses. The DoD has been keen to explore new concepts for advance defense. While they have been discussing active cyber defenses openly, officials are generally reticent to describe what they mean. Privately, they say this would not involve reaching outside of their own networks, even though the defenses were described by the deputy secretary as being "part sensor, part sentry, part sharpshooter."[15]

The Department of Commerce's most important projects have been to implement the new strategy on trusted Internet identities, oversee the Internet domain name system, and assist other nations with capacity development. The State Department, with its new coordinator reporting directly to the secretary, has been increasingly active, especially on issues of Internet freedom, and has taken the lead in bilateral (such as with Russia) and multilateral (in the G8 and OECD) discussions. The Justice Department continues to prosecute criminals and train judges, prosecutors, and law enforcement to recognize cybercrimes and get convictions.

What are possible domestic partnerships that can strengthen national capacities?

Many of the existing "public-private partnerships" primarily serve one of four overlapping purposes:

1. Information sharing (such as for new vulnerabilities or threats). These partnerships include Information Sharing and Analysis Centers (ISACs);

2. Incident response, including the Forum of Incident Response and Security Teams and ISACs;

3. Outreach, such as the FBI's InfraGard partnership; and

4. Policy advice and coordination, like the Financial Services Sector Coordinating Council (FSSCC) and the National Security Telecommunications Advisory Council (NSTAC)

Of course, the private sector is also involved in many other areas of cybersecurity, from education and training to operating and developing core Internet functions, standards, and equipment. Moreover, the DoD's new pilot program with the Defense

Industrial Base is breaking new ground in sharing threat information between the government and DIB and between DIB companies themselves.

To further common goals, the U.S. government should consider several additional kinds of collaboration.

- Response: Looking at the National Cyber Incident Response Plan, one may think that the private sector only has an ancillary role to play. Yet for many kinds of incidents, such as the outbreak of Conficker malicious software, the private sector—not DHS—has been at the center of the national and international response. Accordingly, the government must reconsider its response plans for the many situations in which it will play a supporting, not lead, role.

- Global Norms: Many Western companies and industry associations share goals similar to those of the U.S. government. For example, all want the future Internet to look more like the U.S. version than the Chinese. Yet, to date, the U.S. government has made too few efforts to better encourage the private sector to self-organize and work toward a generally shared vision.

- Environmental Approach: The cybersecurity paradigm may have reached its natural limits, hemmed in by zero-sum concerns over privacy. A new paradigm, seeking not a "secure cyberspace" but a "clean cyberspace environment" may find new ways past this local maximum. Certainly, a U.S. global commitment to work for "clean food, clean water, clean Internet" would not only tie security to the larger development agenda but would be much more likely to enroll young people who care about the environment and have grown up in the digital world.

How does this compare to how other governments are organized to approach the problem?

With the new International Strategy for Cyberspace, the United States has leapt ahead of other governments, by putting forward a complete vision of cyber statecraft, combining security, intellectual property, Internet freedom, and deterrence. We should expect all future national cyber strategies to become similarly encompassing.

Until recently, the United States had one of the most "militarized" approaches to cyberspace, with a strong visible role for the Department of Defense. Though the size and scope of U.S. Cyber Command are still unparalleled internationally, the recent administration strategies have downplayed the military role. Many other nations, though, see cyber commands as the new must-have accessory. South Korea, Germany, the United Kingdom, and Japan have all recently created, or will soon create, new military cyber centers.

The United Kingdom has been very active in broadly similar ways to the United States, with an Office of Cyber Security and Information Assurance (OCSIA) under the Cabinet Office's National Security Secretariat. The OCSIA, along with the Government Communications Headquarters (the equivalent of the National Security Agency), oversee the more technical Cyber Security Operations Centre.

France was perhaps the first Western nation to declare sovereign borders for Internet content, forcing Internet companies in 2000 to respect French laws limiting access to Nazi material. Over time, more nations have been adopting the French model, insisting on some national oversight of content. France has been using its G8 presidency in 2011 to find agreement on the best balance between cyberspace regulation and innovation.

Australia has tackled cybersecurity with a stronger regulatory approach, making a novel distinction between cyber security (concerned with confidentiality, integrity, and availability) and cyber safety (focused on harmful content, such as exposure to illegal and offensive content, cyber-bullying, and stalking). Australia is building up its technical perimeter to keep out such "safety" threats.

To a greater degree since the 2007 large-scale attacks, Estonia has been hitting above its weight, especially with regards to seeking global cyber norms. In its recent cyberstrategy, Estonia set a goal to "achieve worldwide moral condemnation of cyber attacks that affect the functioning of society and impinge directly on people's wellbeing."[16]

Russia has long been active in cyber operations and seems to have both significant capability and strong oversight from its Security Council. Internally, Russian leadership seems to depend more on "scientific" and "technical" experts for what in the United States would be pure policy issues, but there appears to be strong internal and international dialogue.

Unfortunately, China is hyperactive in cyber operations but without similarly strong oversight. Though China has a good interagency process at the mid-levels, there is no clear link for interagency experts to pass information up to their leaders. This also means that those leaders cannot quickly get answers during fast-moving crises, such as in response to questions from Washington, London, or even Moscow. Conflicts and competition involving China in cyberspace will thus only become less transparent and more unstable, and it will become more difficult for opposing sides to signal each other.

What to watch next?

This chapter has described an American cybersecurity apparatus that has recently taken large leaps toward getting its act together. If enacted, these new policies will help the government significantly improve at many basic tasks and start forward on a few more advanced areas. However, weighty problems loom:

1. Limited Action and Scale: The last months have seen an amazing release of policies, but not action. Moreover, many of the initiatives that do exist are little more than pilots that may be difficult to scale up to cover even just the companies in the critical infrastructure sectors.

2. Distraction: With budget battles and looming elections, accomplishing any work done inside the Beltway will be difficult, and there will be sparse funding for new investments or existing initiatives.

3. Lack of Budget Authority: Several important commentators and commissions have called for the White House Cyber Coordinator to have budget authority for more bureaucratic clout. Without this, the interagency process may be effective when there is general consensus but lack teeth to enforce less popular decisions.

4. Mixed Leadership: There are few senior leaders who have a deep understanding of cyber issues and national security, while also being familiar with individual departments and the interagency process. One reason for the recent surge of strong policies has been the presence of strong leaders at the Departments of Homeland Security, Defense, Justice, and (more recently) State. However, DHS has just lost their primary cyber leader (Phil Reitinger, deputy under secretary) while DoD's efforts may become derailed as they have lost three more (Bill Lynn, deputy secretary of defense; General James Cartwright, vice chairman of the Joint Chiefs of Staff; and Bob Butler, deputy assistant secretary of defense for cyber policy).

5. Too Light a Touch? These new policies begin to open the door to new regulation (such as requiring auditors and SEC reporting for critical

infrastructure companies), although today it is still relatively limited. There is a good chance, however, that free market policies have failed and governments will need to adopt stronger enforcement measures, which may be increasingly unlikely in the present political environment.

6. Lack of Measurement and Control: Better security is difficult to attain when it cannot be measured and when control comes at a high cost. Currently, the federal government has a poor understanding of its inventory and has in place only the most rudimentary, often misleading, measurements (such as the FISMA act). In comparison, the best practice companies have a standard baseline that is continually monitored, patched, measured, and reported to senior levels. This decreases the cost of control, allowing more advanced tasks to be undertaken economically. Until the government takes such basic steps to reduce this cost of control, all efforts will be more difficult: Patching vulnerable systems will take too long, as the inventory is not well understood and all hires require long on-the-job training, as every organization has a different set of systems.

7. Changing Technology: The new cyber strategies are a great leap ahead, but they may not have gone far enough. Mobile and cloud technologies are just the most obvious disruptive technologies that will challenge the plodding U.S. policy process. Certainly, there will be more technologies, promising longer-lasting disruptions not far ahead.

With the recent strategies, the United States government has much to be proud of.[17] Optimism is called for, even though it is the optimism of low expectations. As there have been so many failures, it is easy to become excited at just getting by. While there are many more challenges to come, the government has finally shown it can learn lessons and produce strong policies. Implementation will be harder, but at least there are real ends in sight.

Jason Healey is concurrently the Cyber Conflict Education Coordinator and Director of the Washington, DC office of Delta Risk LLC and the Director of the Cyber Statecraft Initiative of the Atlantic Council. He is also a lecturer in cyber policy at Georgetown University and has authored or co-authored papers on cyber conflict for CCSA, the National Resource Council and the Center for a New American Security. From 2003 to 2005, Mr. Healey served as Director for Cyber Infrastructure Protection at the White House. He has worked twice for Goldman Sachs: first from 2001 to 2003 to anchor their team to respond to malicious cyber incidents and later, as an executive director in Hong Kong from 2006 to 2009, to manage Asia-wide business continuity and create the bank's regional crisis management capabilities to respond to earthquakes, tsunamis, or terrorist attacks. Mr. Healey is a former Vice Chairman of a key public-private homeland security partnership, the Financial Services Information Sharing and Analysis Center. Starting his career in the United States Air Force, he had intelligence and information warfare tours at the National Security Agency and Headquarters Air Force at the Pentagon and is a plankholder (founding member) of the Joint Task Force – Computer Network Defense, the world's first joint cyber warfighting unit. Mr. Healey holds degrees from the United States Air Force Academy (Political Science), Johns Hopkins University (Liberal Arts) and James Madison University (Information Security).

[1] TIME magazine seems to have had the first major cover story of "Cyber War" on 21 August 1995, nearly 16 years before the latest major cover story from Bloomberg Businessweek, which declared on 25 July 2011 that the "Cyber War Has Begun."

[2] The list of wake-up calls include the Morris Worm and Cuckoo's Egg intrusion (late 1980s), the SOLAR SUNRISE intrusion and exercise ELIGIBLE RECEIVER (late 1990s), MOONLIGHT MAZE intrusions (circa 2000), Chinese espionage intrusions (early 2000s to the present), attacks against Estonia and Georgia (2007 and 2008), and BUCKSHOT YANKEE intrusions (2008). Note this list does not even include intrusions into Google, stolen F-35 information, WikiLeaks, or other recent newsworthy intrusions.

[3] The 1997 release was based on the Marsh Commission report and led to President Clinton's PDD-63 and the establishment of new organizations at the Federal Bureau of Investigation and Department of Commerce. The 2003 release was part of the overall focus on homeland security after the 9/11 attacks and included HSPD-7, NSPD-38, the National Strategy to Secure Cyberspace, and the National Infrastructure Protection Plan and accompanying organizations, especially within the Department of Homeland Security. The Comprehensive National Cybersecurity Initiative of 2009 remains important, but was still limited to government actions and networks.

[4] Indeed, this author, while working as a policy director in the White House office that published the 2003 strategy, was told to ignore it less than a year after it was published.

[5] Though many observers thought it would itself be the new strategy, this was only #2 on the list of near-term actions.

[6] Taken from the fact sheet on the new International Strategy for Cyberspace, emphasis added.

[7] Released on 12 May 2011 and available at http://www.whitehouse.gov/blog/2011/05/12/administration-unveils-its-cybersecurity-legislative-proposal.

[8] Ibid.

[9] Released on 8 June 2011 and available at http://www.commerce.gov/news/press-releases/2011/06/08/commerce-department-proposes-new-policy-framework-strengthen-cybersec.

[10] Released on 13 October and available at http://www.sec.gov/divisions/corpfin/guidance/cfguidance-topic2.htm.

[11] Released on 16 May 2011 and available at http://www.whitehouse.gov/blog/2011/05/16/launching-us-international-strategy-cyberspace.

[12] Released on 14 July 2011 and available at http://www.defense.gov/home/features/2011/0411_cyberstrategy/.

[13] Conversation with author, 15 March 2011.

[14] William J. Lynn III, "A DoD Cyberstrategy To Prepare Our Military for Emerging Cyber Threats" (speech delivered at National Defense University, Fort McNair, Washington, DC, July 14, 2011).

[15] Lynn, "Defending a New Domain: The Pentagon's Cyberstrategy," *Foreign Affairs*, September/October (2010).

[16] Released in 2008 and available at http://www.mod.gov.ee/files/kmin/img/files/Kuberjulgeoleku_strateegia_2008-2013_ENG.pdf.

[17] This is the first time in a long time this author has felt jealous of White House staffers, who have penned an extraordinary, brilliant international strategy and shepherded it through a long interagency marathon.

"*The fundamental issue in question is the role of the state. Given the transnational and cross-border nature of malicious activity, security requires more than action at the enterprise and national levels. International cooperation is essential. Nation-states are more capable and experienced in cooperating on security than civil society, which lacks the mechanisms, authority, and legitimacy to be effective.*"

—JAMES A. LEWIS

Harnessing Leviathan:
Internet Governance and Cybersecurity

James A. Lewis
Senior Fellow and Director
Technology and Public Policy Program
Center for Strategic and International Studies

Cybersecurity has attracted considerable attention, but this has not improved our understanding of the problem. Discussion is hampered by imprecise terms, jargon, exaggeration, a lack of data (and a consequent reliance on anecdote and fable), business interests, and ideology. The debate includes a range of communities who have limited experience in military or international affairs, but are stoutly convinced of the centrality of their own insights and expertise. Larger trends in American political thought also distort analysis. The call to shrink government and rely on the private sector and markets to address public problems contributes to weak cybersecurity—a government, in the infamous words of Grover Norquist, "small enough to drown in a bathtub" is no match for advanced foreign opponents.

Cybersecurity can mean the safeguarding of individual networks and the data resident on them from malicious activity – this limited definition was adequate when the Internet was small and unimportant. It can mean the security of a new domain, allowing individuals and companies to act without undue risk of harm. But cybersecurity must also mean the preservation of the essential values that have guided American foreign policy. An outcome where the Internet becomes more secure but less free would be a setback for the U.S.

Governance—the process of creating or amending rules and the mechanisms to secure compliance with them—is cybersecurity's fundamental problem. The original American view was that Internet governance should be weak and the role of government strictly proscribed, as this would empower innovation and allow an emerging global community to guide the new infrastructure. Security was largely ignored. As a result, the Internet has become an unparalleled vehicle for espionage and crime, a true Hobbesian environment. Nations are increasingly concerned about cybersecurity and want international action to limit risk. The failure of the current governance structure to provide security creates powerful and unavoidable tension that has led other governments to seek an expanded role in cyberspace. There is an emerging international consensus that cyberspace must be governed like other global activities, by a web of relations and commitments among nation-states. The risk in moving to a new approach towards governance, however, is that it creates an opportunity to extend sovereign control of the Internet in ways that diminish key democratic values.

Many countries believe that the U.S. has entrusted management of the Internet to a nonprofit corporation in California, whose behavior has been erratic and whose ties to the U.S. government are murky. They would prefer a management structure directed by states. Those who oppose this solution proffer alternatives that rely on non-state actors, but these alternatives have not worked. There is neither a technological silver bullet nor any informal coalition of "civil society" Internet users that can succeed against the malicious actors who take advantage of global connectivity, porous networks, and weak governance.

The U.S. has fought a slow, rearguard action to block change in Internet governance, but frustration is growing and other states are seeking ways to circumvent or overpower U.S. objections. The fundamental issue in question is the role of the state. Given the transnational and cross-border nature of malicious activity, security requires more than action at the enterprise and national levels. International cooperation is essential. Nation-states are more capable and experienced in cooperating on security than civil society, which lacks the mechanisms, authority, and legitimacy to be effective.

That better governance will produce better security and that better governance requires a predominant role for national governments are by no means uncontested assertions. But the failure of the non-governmental approach to secure cyberspace lies heavily over the proponents of a civil society, market-driven governance. One test of the assertion that governance by nation states is essential to cybersecurity is to ask if there are realistic alternatives. While it is tempting to list all the flaws of a state-based approach—the slowness, the discord, the competition—the alternatives are demonstrably worse. No international system involving states will be perfect, but other models for international cybersecurity, which generally ascribe to beliefs that better technology will save us or that civil society and markets can provide adequate security, are even more flawed. In no area of international security – proliferation, terrorism, regional stability, arms control – would we accept the premise that voluntary action by private individuals is sufficient. And yet, despite its inadequacy, this notion has dominated the cybersecurity debate. It should now be abandoned.

The politics of a new approach to governance are complex. The end of the Cold War did not mark an end to conflict and competition among states. The U.S. now finds itself in a world where alliances are less cohesive and allies less powerful. But the last decade has seen the rise of powerful new economies—Brazil, India, China—who challenge the U.S. for international influence and regional leadership. Among these new powers there is dissatisfaction with the international institutions assembled after the Second World War, with their transatlantic focus and deference to Europe. The new powers believe to varying degrees that the U.S. designed the post-1945 international order to provide itself with an economic and military advantage. They want to change this to reduce the U.S. advantage and gain it for themselves. Although aspects of the current situation resemble earlier multipolar episodes, when the great powers competed, we are not yet in the multipolar environment of 1900, with competing alliances that must be counterbalanced for security. Moreover, the terms of competition are now different: the powers are competing for influence over the structures and rules of global finance and business, rather than for colonies and resources.

So far, this challenge is a reaction to American power rather than an effort to replace it. There is yet no coherent alternative to the conceptual framework for international order assembled by the U.S. and its allies after World War II. None of the new powers has an alternative vision as to how the world should work, only a belief that what has been inherited is inadequate because they were not involved in its creation and that they should play a greater role in its management. This belief, in turn, shapes their views on Internet governance and cybersecurity.

The decline—perhaps temporary—of U.S. global influence will also shape any effort to secure cyberspace. Influence is measured by the ability to secure a desired outcome; by this measure, the U.S. is weaker than it was ten years ago. There are several contributing factors: a series of missteps in the first decade of this century and a belief that U.S. policy was largely responsible for the global recession. The result has been the creation of powerful "antibodies," where if the U.S. is for something, other influential nations are automatically suspicious, if not opposed to it. A nimble foreign policy could circumvent and exploit these antibodies and gain support from the new powers, but the U.S. lacks a strategy tailored to take advantage of the new circumstances.

The Council of Europe Cybercrime Convention, a comprehensive agreement that provides a strong legal framework for cooperation, illustrates the problem. The Convention has effectively been blocked. Some nations—Russia, for example—

object to the Convention, noting that signing it would allow other countries' police forces to violate the sovereignty of member states. This argument is specious, and masks a deeper reluctance to work against cybercrime. Other nations, such as Brazil and China, complain that they were not involved in negotiating the Convention and refuse to sign a document that they did not help draft. The Cybercrime Convention now faces competing agreements tailored to attract developing countries. The most important of these is the Shanghai Cooperation Organization (SCO), a loose coalition of countries who cooperate on security. The SCO is presented as an alternative to the Council's Cybercrime Convention, but its rules are vague and its membership restricted. In addition to the SCO, Russia and China have pushed the International Telecommunication Union (ITU) to play a larger role in cybersecurity, an effort that has made some progress. Neither the SCO nor the ITU is particularly effective, but they are alternatives to the transatlantic approach.

American interests include, as they have for a century, the promotion of a stable international order based on the rule of law, open and equitable arrangements for trade, and a commitment to democratic government and human rights. While the U.S. record is not perfect, no other nation has as consistently or forcefully pursued these ideals, and no new competitor has the same commitment to them. This means that the creation of alternate and competing governance structures for the Internet and cybersecurity could undercut America's long-term interest in a stable and secure international order. Avoiding this outcome will require cooperation with both allies and emerging powers to create a collective approach to cybersecurity based on norms, laws, and institutions.

Misconceptions Damage Security

We lack a clear understanding of the fundamental nature and problems of cybersecurity. Cybersecurity requires identifying which instruments of international order are now necessary, where the extension of existing governance structures into cyberspace is sufficient, and where the creation of new institutions, norms, and laws is needed. A useful first step in this process of identification is to reduce the surrounding cacophony. Cybersecurity is not much different from any other issue in international security; the same political and economic forces shape it. The specifics of the technology affect both problems and solutions, just as they do in trade, nonproliferation, and arms control. There are areas of ambiguity, but cybersecurity is neither sui generis nor subject to such rapid change that intervention is impossible.

An emphasis on the speed of technology, and how this limits the scope for government action, is a rhetorical device that we can discard.

Our most dangerous opponents are the military and intelligence services of other nations and their proxy forces. The counterfactual that illustrates the problem is that if Russia and China ended their cyber programs and no longer tolerated cybercriminals on their territories, the scope and sophistication of exploits against the U.S. would decrease significantly. These opponents have the resources and commitment to overcome most defensive efforts, particularly the disaggregated, voluntary defense used by the United States. The ideas underpinning American cybersecurity—public private partnership, voluntary action, information sharing—date to the 1990s and are now slogans rather than policies. Cybersecurity that relies on voluntary, disaggregated action will always be inadequate against state opponents.

Although malicious action in cyberspace is constant, there have been only two real cyberattacks. The UN Charter, The Hague and the Geneva Conventions make clear that an attack involves physical destruction and casualties. Only Stuxnet and the Israeli air raid on the alleged Syrian nuclear facility can therefore be considered attacks. The distributed denial of service incidents in Estonia and Georgia were not attacks, although they do raise important questions about the nature of cyberwar. A non-destructive event like the one in Estonia, if it were of greater scope and duration and blocked key services for an extended period, might qualify as an "attack." Similarly, a massive erasure of data might be judged equivalent to physical damage. These are areas of ambiguity, but calling everything an attack is inaccurate and unnecessarily complicates the discussion. A precise definition of attack also refines policy options on the use of military force to deter cyber exploits. As espionage and state-sponsored crime do not qualify as attacks and are not casus belli, the ability to deter them is limited.

Cyberattack on its own, however, will not win a conflict, particularly against a large and powerful opponent. It is not a "decisive" weapon, but a new military capability that combines global reach and high speed with a payload that is less destructive than kinetic weapons. Militaries will use cyberattack to complement other capabilities: The immediate goal of a cyberattack will be to create confusion and uncertainty among opposing commanders, by attacking networks and data. Cyberattacks may also damage or destroy critical infrastructure, but an attack that destroys civilian targets in the opponent's homeland could well be considered escalatory. Fear of escalation may curb the use of cyberattacks on infrastructure, at least in the opening phases of conflict.

The problem of attribution is also overstated. Most analyses are inadequate as they are based on a discussion of forensic techniques and private sector experience. Identifying the author of a single event is difficult, but with multiple events the likelihood of attribution increases. Attribution can be reinforced by active intelligence measures, using methods not available to private actors. The need for attribution varies by scenario: Attribution is a problem for law enforcement, with its high evidentiary standards, but it is less of a problem in military conflict. There is always uncertainty in warfare, requiring judgments by commanders and policymakers. Similarly, uncertainty is normal in espionage (by its nature covert), and decisions must be made using slender, incomplete, or unreliable information. Cyberwarfare will be no different.

Cybersecurity is not simply a technical issue. A technical approach may have been adequate for the first years of the Internet, but cybersecurity now requires the resolution of key political questions on the role of government and the nature of conflict and competition among states. Nor is cyberspace a commons. It is a man-made construct that depends entirely upon a collection of interconnected devices, all with individual owners and all subject to sovereign control. The extent of sovereign control was initially obscured by the intoxicating expansion of connectivity among Internet users and by a belief that globalization would make borders irrelevant. Nations have since discovered that borders exist in cyberspace and that they can exert sovereign control within them. A better description would be cyberspace as a condominium with many contiguous owners, whose shared interests are administered by a diffuse collection of weak governing bodies. The United States deliberately put in place this loose governance structure when it commercialized the Internet. It reflects the ideals of the pioneers of cyberspace, but it is inadequate for security, particularly as the Internet spreads to nations with very different values and laws. There are few rules and no adequate processes to develop them. Defining cyberspace as a condominium highlights the need for new governance structures to make cyberspace more secure.

A Heritage of Weak Governance

Internet governance is an artifact of the politics of an earlier American era. When the Internet was an American entity and the new technologies had an aspect of complexity if not magic, nations deferred to American views. When the Americans asserted that cyberspace was a commons, that technology would outpace government action, that innovation required an unconstrained environment, and that governance should derive from some natural community of civil society stakeholders, other nations accepted this. Now, the old deference to the American vision has eroded, and there is dissonance between American views and those of other nations. Many would prefer the UN and its agencies govern the Internet. The struggles over the International Telecommunication Union (ITU) or the Internet Corporation for Assigned Names and Numbers (ICANN) reflect this dissonance and present a growing challenge.

What we have now is a collection of inchoate governance entities—none particularly useful. The best-known entity is ICANN, which maintains the Domain Name System, the Internet's "address book." ICANN and the Internet Engineering Task Force, which oversees the development of technical standards, were created by the U.S. to "manage" the Internet in a technical sense. And while they are effective at their technical tasks, they are not governance bodies. As nonprofit organizations outside of the control of national governments, they attract criticism from many nations. Other Internet governance entities, like the UN's Internet Governance Forum, are intentionally feeble, created to give an outlet to the frustrations of other nations without providing any ability to make changes.

Proposed alternatives for governance carry political and perhaps commercial risk for the U.S. and its allies. The leading alternative to the American system is the ITU, a UN body that manages spectrum allocation and telecommunication standards (one of its primary functions is to assign international calling codes). As a UN body, it is more amenable to the interests of governments. Countries that want governments to play a larger role in controlling cyberspace and its technologies use the ITU to undercut the multi-stakeholder, private sector-led approach to governance preferred by the United States. But a larger role for the ITU could come at the expense of cybersecurity, by not adequately addressing the fundamental political problems of Internet governance. The ITU does not have the expertise or jurisdiction to develop policy, rules, and governance for cybersecurity (much less for law enforcement or armed conflict). Nevertheless, because it is a more malleable body, rival states like Russia and China have pushed the ITU to become the de facto center for cooperation and capacity building in cybersecurity, an effort that made some progress during the period of official U.S. disinterest in the international aspects of cybersecurity.

Although the United Stated is not used to seeing global initiatives begin without it playing a leading role in their development, this is what is happening in cyberspace. The current governance structure is untenable; proven to be inadequate for managing and securing a global infrastructure. Continuing to support it undercuts U.S. leadership and puts U.S. interests at risk. Reassessing U.S. policy is essential, and this means asking what international rules and institutions will make cyberspace more secure.

Building a Cooperative Framework for Security

The minimum requirement for cybersecurity is the concurrence of nations that they will observe existing international commitments in trade, security, finance, and law enforcement in cyberspace. We do not yet have this concurrence. A bank robber should not gain immunity by using the Internet to commit transborder crimes. Intellectual property should not be even less protected when copying and piracy are carried out over networks. The norms governing the use of force in international relations and the laws of armed conflict—proportionality, distinction, discrimination in attack, and others—should equally apply for cyberattack and cyberwar.

In part, nations have been slow to extend commitments because of the belief that cyberspace is a new domain where the rules of the game do not apply. More importantly, there is little incentive to extend commitments since there is little or no consequence for malicious action in cyberspace. Instead, we allow individuals and nations to safely ignore their obligations in cyberspace. If the only cybercriminals ever convicted are those foolish enough to live in the West—while those who reside in Russia face neither arrest nor indictment—we should not be surprised that cybercrime is a growth industry. If Chinese hackers are permitted, if not encouraged, to engage in economic espionage and China need not fear any retaliation, opportunity seems unbounded.

Agreement among nations that existing obligations and responsibilities apply to cyberspace and the development of penalties for malicious action would begin to make cyberspace more secure. This agreement would also help identify those areas where new rules, norms and institutions in how states relate to each other in cyberspace are required. Some argue that bringing law and order to cyberspace will close off opportunities for innovation, and that the cost of poor cybersecurity is offset by the gains to innovative capabilities. Perhaps this was true in the early years of the Internet, but now that the global economy and international security depend on it, we can no longer afford inadequate security. In any case, the asserted link between an unconstrained Internet and innovation, although an article of faith, is dubious and based on questionable assumptions about economic change, entrepreneurship, and technology.

The development of norms—shared expectations about behavior—may be the most important governance requirement for cyberspace. Norms can be explicit or implicit; they shape behavior and limit and shape conflict and competition. There are currently only a few implicit norms in cyberspace, the most important being an unspoken tenet among nations to avoid cyber actions that could be interpreted as the use of force. While this provides an uneasy stability, it is inadequate, if only because avoiding the use of force does not preclude the commission of crime or espionage. We need additional and explicit norms to effectively govern cyberspace. By examining the greatest sources of risk, we can derive four general principles to guide cybersecurity internationally:

1. States agree to cooperate to proscribe malicious practices in cyberspace.

2. Existing international commitments and law, such as the laws of armed conflict, apply in cyberspace.

3. States are responsible for the actions in cyberspace of those resident in their territory.

4. States exercise restraint in supporting the acquisition of cyberattack or cybercrime capabilities, and prevent acquisition by terrorists, criminals, or other dangerous groups.

Norms based on these principles would make it difficult for states to deny responsibility for malicious action. It could be argued that such norms are already incumbent upon states, but key nations do not accept this for cyberspace, and there has been no effort to hold them accountable. Acceptance of these norms would pose challenges for China and Russia, both of which make extensive use of proxy forces (as in the case of Estonia). It would also pose challenges for the U.S., where lax policies allow unwitting American consumers to become a major source of cybercrime when cybercriminals surreptitiously capture computers and make them a part of "botnets" used for criminal purposes.

Restraint in providing attack capabilities to worrisome groups parallels other regimes and norms limiting the transfer of weapons. As with nonproliferation guidelines, the intent is not to restrict access to technology, but to avoid transfers that put international security at risk. Implementation would require judgment on the part of supplier nations as to their own interests and responsibilities and to the intent of the recipient. A ban on cyberattack capabilities makes little sense, given the pervasiveness of the technologies and its many civil applications.

Russia and China argue that the existing laws of armed conflict are inadequate for cyberwar. Discussions with Russian and Chinese officials indicate they believe that information is a weapon that Americans will use against them to destabilize their governments. It is the desire to control information for political purposes that leads them to argue that existing laws are inadequate. We can reject this, however, as it dissociates attack from the use of force. The use of force produces physical damage; supplying alternate sources of information does not. Treating information as a "weapon" runs contrary to the long-standing definition of the use of armed force that guides international agreements.

In 1998, At the UN, Russia put forward an unverifiable treaty that would have banned certain kinds of cyber "weapons" and attacks. Many nations initially supported the treaty out of concern over U.S. proclamations that it would "dominate" cyberspace and because there was no viable alternative. Although the Russian treaty was too flawed to be adopted, in light of international concern the Secretary General organized a Group of Government Experts to examine the prospects for international cooperation in cybersecurity. The group was tasked with considering how existing laws of conflict apply to cyberwar, and how (reflecting the concerns of China and Russia) sovereignty and national control affect information entering a country over its networks.

Prior to 2010, the U.S. multilateral engagement on cybersecurity had confined itself to a sterile exchange of best practices. The new U.S. government international strategy for cyberspace (issued in May 2011), however, places engagement and the development of norms at the center of U.S policy. After years of blocking any serious international discussion of cybersecurity, the U.S. made an abrupt about-face in 2010, when it led the effort in the UN to secure agreement on an initial framework for international cooperation. The July 2010 report issued by a group of 15 government experts (and later endorsed by the General Assembly) showed a high degree of international concern about the risk of cyber conflict and a general agreement on the need for international cooperation to limit its risk. Unfortunately, the report also demonstrated that while most states share concern about the risks of cyber, key nations—in particular the U.S., Russia and China—differ on the specifics of any international agreement.

The approach endorsed in the UN highlights norms, transparency, and cooperation rather than putting in place a "cybersecurity treaty." Binding formal agreements to ban or restrict technology do not make sense when the technology is widely available from commercial sources, is in widespread use, and when verification of compliance is impossible without unacceptably intrusive measures. No advanced nation is likely to renounce cyberespionage. Precedential UN efforts are of limited value: Since cyberattack does not entail the same degree of horror and destruction as weapons of mass destruction, and since malicious cyber techniques, unlike WMD, are regularly used by many nations, a treaty modeled on the Chemical Weapons Convention or Biological Weapons Convention would lack the political force necessary for adoption. If a treaty approach were pursued, the outcome would be feckless, or, at best, make only imperceptible progress, because any treaty would be unverifiable and "weaponizable" technology is easily accessed.

In the Cold War, the U.S. and the Soviets came to understand implicitly that nuclear weapons would be used only in extremis. Nuclear weapons were, in a sense, stigmatized. Doctrinal debates and military and political exchanges allowed each side to better assess intentions and risk. Similar transparency for cyber conflict could also reduce risk. The nuclear precedent is limited, however. The horror of nuclear war made its use politically unacceptable; the same is not true for cyber weapons. While most cyberattack doctrine remains secret, some elements have emerged that suggest that cyberattack forms an important part of planning for military conflict. For any agreement to be meaningful, it must take into account the reality that cyber weapons will, in fact, be used.

Cybersecurity is more likely to be increased using confidence building and transparency, to reduce the chances of misperception or misinterpretation of actions in cyberspace, and by establishing common understandings on cyberwarfare that reduce the chance of inadvertent conflict. Greater insight into how combatants will decide when and how to use cyberattack would be stabilizing. Building confidence through greater transparency in cyber doctrine, in either bilateral or multilateral exchanges, could reduce the chance of miscalculation or inadvertent escalation. Miscalculation or misinterpretation could arise because cyberattack and espionage are almost identical. In both cases, the attacker probes the target network, gains access, and implants malicious software. Transparency could reduce misperception, but it will be necessary to engage potential opponents, as was the case for strategic arms control during the Cold War, to achieve this – and engagement must be more than an exchange of press releases.

Bilateral discussions with Russia and China in the last year have had initial success in reducing misperceptions and identifying areas for future agreement. These have included exchanges of information on defense planning, discussion of thresholds for what qualifies as an attack in cyberspace, and the applicability of norms for conflict.

Chinese proposals include the idea of "no first use" and creating sanctuaries—civilian targets that would be excluded from cyber strikes. Both proposals have problems, but they raise the question of whether it would be possible to "stigmatize" certain kinds of cyberattacks against civilian targets. While progress has been made in both sets of bilateral exchanges, both are at a very early phase and much work remains to be done. Progress will require U.S. leadership and engagement.

That cyberspace is a global network does not mean that a single approach to cooperation is best. This is an essential point, as it removes a frequent objection to creating cyber governance structures—that it is premature to develop rules for a protean technology that continues to evolve. It would indeed be an error to develop a single, overarching governance structure, but it is more than past time to create specialized instruments for those areas where governance is weak, or to determine how to extend existing instruments and commitments into cyberspace. An effort to develop an agreement that addresses all the problems cybersecurity entails—crime, espionage, warfare, trade—will be hopelessly complicated, but this is not the only option. There are linkages, of course, among these problems, and these linkages complicate multilateral efforts, but cybersecurity will require multiple, specialized mechanisms for law enforcement, military activities, trade, and governance. Some of these bodies already exist while others will have to be created.

Alternative governance structures could be modeled on the Missile Technology Control Regime (MTCR), the Financial Action Task Force (FATF), the International Civil Aviation Organization (ICAO), or even Bretton Woods. These regimes use norms, rules, and compliance mechanisms to create a cooperative approach to security. The FATF, for example, is a group of central banks and law enforcement agencies that came together to work against money laundering by requiring a nation's banks, if they wish to participate in the global financial network, to observe FATF rules. If they do not, their transactions may be slower, more complicated, and more expensive. The UN, led by the U.S., created the ICAO at the end of World War II to set standards to ensure that the airlines of one nation could operate safely in another, on rules for passengers, and on accident investigations. The ICAO provides the governance structure that enables the global air transport network to operate safely. Similar governance structures for cyberspace could proscribe harmful practices and develop cooperative measures to improve security.

An effective international cyber strategy will require the U.S. to simultaneously engage with different groups—in the UN or G-20, with like-minded nations (NATO or perhaps the OSCE), and with potential opponents. At a minimum, the U.S. must undertake initiatives in the UN, if only for defensive reasons; we do not wish to cede this forum to potential opponents. Those who do not share our values will try to use the UN to capture the initiative on cybersecurity. Progress will be slow, and the U.S. may need to test different questions on what issues are best dealt within the UN, which should be addressed in other international fora, what problems can be mitigated by the extension of existing agreements, and which require new understandings and commitments. The U.S. may need to employ a multi-pronged effort similar to that used to build international consensus on nonproliferation in the early 1990s.

Cybersecurity and the State

Early thinking about governance was shaped by the belief that the Internet would provide the means for a new kind of governance: Cyberspace would be a global commons managed by a self-organizing community. This reflects widely held views from the 1990s on the future of international relations: borders would be less important, governments would see their role shrink as civil society (multi-national corporations and NGOs) assumed greater responsibility, and the Westphalian state system would be weakened and perhaps replaced. At best, these concepts of a post-Westphalian future can be described as premature, if not utopian.

The extension of sovereign control into cyberspace is the single most important trend shaping governance and security. The move toward a new governance structure for cyberspace brings with it real risk. Other nations will try to use this evolution to control technology and content for political and commercial purposes in ways that could harm U.S. interests and security. The challenge for the U.S. is how to manage this transition from the Internet's pioneering governance system to a structure that provides greater security while preserving democratic values.

Answering this challenge will require the U.S. and other nations to make fundamental decisions on governance, the most important of which is to determine the values and rules that will shape the extension of national sovereignty into cyberspace. If the U.S. does not provide a new vision and a framework for governance, it is quite possible that a new approach will not reflect the principles of openness and free expression we have come to think of as inherent elements of cyberspace. Other nations will attempt to restructure cyberspace in ways that may not serve the global good or our national interest. The pressure to secure cyberspace creates an unavoidable need for change. The question is now who will lead it and what principles will guide it.

The political challenge of cybersecurity comes at a time when Western influence on global institutions is waning. The U.S. could once dictate the course of the Internet, but this is no longer the case. Just as the U.S. and its allies created rules and institutions for global activities in finance, telecommunications, trade, and air travel, the same must now be done for cyberspace; but the requirements for governance are still ill-defined and in dispute. There are contending visions for cybersecurity, and areas of deep disagreement. Unilateral domestic measures are inadequate. Only international agreement can make cyberspace more secure. Progress is not impossible, but these are complex issues, and serious improvement will require both a difficult reappraisal of policy and a sustained effort at engagement.

James Andrew Lewis is a Senior Fellow and Program Director at the Center for Strategic and International Studies, where he writes on technology, security, and the international economy. His current research involves the political effect of the Internet, innovation and economic change, and strategic competition. Before joining CSIS, he served at the Departments of State and Commerce as a Foreign Service Officer and as a member of the Senior Executive Service. Dr. Lewis has authored more than seventy publications since coming to CSIS and was the Director of CSIS's Commission on Cybersecurity for the 44th Presidency, whose report has been downloaded more than 50,000 times. He was also the Rapporteur for the UN's 2010 Group of Government Experts on Information Security. Dr. Lewis received a Ph.D. from the University of Chicago.

Part **3**

"Network-enabled intellectual property theft and commercial espionage threaten to undermine our national competitive advantage. Ironically, an over-energetic regulatory or bureaucratic response could be equally damaging, by constraining future web-enabled economic gains. Rather, a middle ground is needed, where government stimulates the private sector to protect its most valuable assets."

—JOHN DOWDY

The Cybersecurity Threat to U.S. Growth and Prosperity

John Dowdy
Director
McKinsey & Company

> *"...we must remember that cyber crime, cyber terrorism, cyber espionage or cyber war are simply crime, terrorism, espionage or war by other means. Cyberspace adds a new dimension, but its use in warfare should be subject to the same strategic and tactical thought as existing means."*
>
> — UK Minister of State for the Armed Forces Nick Harvey
> in *The Guardian* ("Forget a cyber Maginot line," 30 May 2011)

Introduction

Cybersecurity has attracted a considerable amount of attention recently, due to a spate of attacks on high-profile government and business targets including the CIA, Sony, Lockheed Martin and Citigroup. Internationally, both governments and corporations are beginning to recognize the scale of this cybersecurity challenge. President Obama launched a legislative proposal to tackle the challenge following the release of a recently concluded policy review, which suggested that "threats to cyberspace pose one of the most serious economic and national security challenges of the 21st century for the United States and our allies."[1]

This article explains why addressing the cybersecurity threat is critically important for U.S. economic prosperity and why the "same strategic and tactical thought"[2] will be ineffective. Government must realize that in addition to the shift it is making from traditional physical security era approaches and mindsets, it must also make a shift to recognize that it is responsible not only for the protection of its own assets, but for cybersecurity in the private sector, as well. The need for change is not limited to government: The private sector must also recognize the severity of the threat it faces and collaborate with government and cybersecurity vendors to address it.

The stakes are high. Cyberattacks seriously challenge U.S. competitiveness by threatening two of the core drivers of U.S. economic prosperity: intellectual property (owned by both government and the private sector) and the Internet. Both government and business leaders need to respond to the threat. Government is already investing in defending its own assets, but it cannot afford to stand aside with regard to assets held by the private sector for two reasons. First, some areas of the private sector are important in government's own supply chain; second, the private sector's intellectual property is vital for economic prosperity.

Unfortunately, our research suggests that while the private sector has significant economic value at risk from intellectual property theft, neither the high value of this intellectual property, nor its susceptibility to cyberattack is fully appreciated. Businesses tend not to prioritize cybersecurity, and the government is doing less to help businesses protect their intellectual property than it is doing to help protect critical national infrastructure or its own classified information. One reason for this may be that while the security agencies have a good understanding of the extent of the threat, this understanding has not been fully absorbed in other areas of government.

In order to address this threat without acting so drastically as to compromise the Internet's contribution to the U.S. economy, the government needs to promote the emerging "security-economic complex," a system with the potential to boost cyber defense capabilities much as the military-industrial complex boosted physical defense. Four key elements of this approach are: (1) Embracing government's responsibility to support the protection of both its own intellectual property and that of private enterprises; (2) Providing incentives to private enterprise and to cybersecurity vendors to encourage enterprises to adopt a more robust approach to the threats they face and incentivize vendors to increase their investment in research and development (R&D). In such an environment, private enterprise and cyber vendors can work together with government to bring about more effective technical and managerial security solutions; (3) Providing private enterprise with enough information and knowledge transfer on the extent and nature of the threat so that companies understand what they are up against; and (4) Establishing a framework within which companies can share details of the attacks that they have faced in order to help prevent future attacks.

The advantage currently lies with cyberattackers. As a result, if the government chooses not to act, the number of attacks will continue to increase—as growing online economic activity and data storage increase the incentive of the attackers—and U.S. competitiveness will suffer.

Cyber Threat is Poorly Understood

Public understanding of the extent of the threat from cyberattacks is poor because data on cyberattacks is scarce. It is very difficult to get a good picture of the real extent and cost of cyberattacks. What, for example, was the true cost of the 2007 attacks on Estonian websites, including those of the Estonian Parliament, banks, ministries, newspapers, and broadcasters? Or of the similar attacks on Georgia in 2008? What was the impact of the alleged loss of data relating to the F-35 Joint Strike Fighter, or the attacks on Sony's PlayStation Network and EMC Corp.'s RSA unit? What was the cost to the U.S. government of the release of its data by WikiLeaks?

Definitive public figures are very hard to come by for two reasons. First, although there are government agencies (such as the National Security Agency) that systematically monitor cyberattacks across the U.S. and hence have a good understanding of the extent of the threat, these organizations do not readily share their knowledge for the sake of protecting their sources and working methods. As a result, neither the public nor many areas of government outside defense and security share an understanding of the extent of the cyber threat.[3] Second, private enterprise and government bodies often do not publicly report the attacks they experience: They have little incentive to do so,[4] and the wide variation in reporting requirements by jurisdiction allows them not to report a breach. According to Dmitri Alperovitch, a cybersecurity expert at McAfee, less than 1 percent of cyberattacks discovered by the target are reported.[5] Moreover, it is difficult to quantify even the exact costs of attacks that are publicly acknowledged. Sony, for example, has announced that it expects its recent data loss will cost the company $173 million,[6] but others have estimated costs of up to $1.5 billion.[7] And the costs of security breaches are not contained to the company alone: Sony's share price fell by 7 percent and Lockheed Martin's fell by 4 percent in the days following their attacks,[8] with resultant losses to shareholders of $2.2 billion and $1.0 billion, respectively. In contrast, the share price of EMC rose in the days following the attack on its subsidiary, RSA. Thus market response would not appear to be a reliable indicator of the cost of cyberattack.[9]

Even fewer figures are available for the economic cost of attacks on government. At the extreme, the alleged attacks on the F-35 program could, by revealing technical specifications to other countries' armed forces, compromise the U.S. government's estimated $285 billion development cost.[10] It is hard even to guess the cost of disruptions in Estonia and Georgia.

The problem is compounded when cyberattacks are not immediately recognized. We have seen examples of companies that have not discovered attacks until after their systems have been breached for considerable periods of time. The "Kneber bot" attack, for example, began in 2008 and was only discovered in 2010, after breaching more than 75,000 computer systems.[11] Operation Shady RAT, revealed by McAfee, involved attacks on more than seventy organizations spanning five years.[12] And some organizations may not be aware of a breach at all. Alperovitch, who authored the McAfee report on Operation Shady RAT, observed that "There are only two types of companies—those that know they've been compromised, and those that don't know."[13] It seems reasonable to assume that there are numerous undetected, possibly significant attacks currently underway.

Overall, extrapolating these different kinds of events into economy-wide figures is problematic. How many attacks of each magnitude occur, and with what regularity? According to a 2011 survey, more than 80 percent of critical infrastructure providers reported being the victims of large-scale cyberattacks or infiltrations—but at what cost?[14] And what number should we assign to the many incidents that are detected but unreported?

This difficulty in estimating the true cost of cyberattacks has led a number of organisations to develop top-down estimates of the scale of the issue that rely on questionable assumptions, yielding implausible figures from which no government can reliably set policy.[15]

Assessing Cyber Threats

The key threats are to critical national infrastructure, the government's classified information, and the intellectual property of private enterprise. To date, neither the government nor private enterprise has acted sufficiently to protect intellectual property. As government (outside those areas dealing with defense and security) and business are unable to accurately determine the cost of attacks, they pay insufficient attention to the value they have at risk of cyberattack and their vulnerability to such attacks. Ignorant of the facts, they are unable to prioritize how they will respond to the most serious threats.

In order to help companies and government estimate how much of their value is at risk, our team at McKinsey examined—for the whole range of attackers and targets—the capabilities of attackers, the vulnerability of targets, and the value at stake in each case. As cyberattackers are not motivated exclusively by money, we also considered whether attackers had non-monetary incentives to attack each target.

This gave us a view of both the likelihood and impact of attacks for each combination of attacker and target. Our most striking finding was the level of threat against private enterprise. In particular, of all assets, the intellectual property of private enterprises has the highest value at risk of attack. This was recognized in a recent speech by Deputy Secretary of Defense William J. Lynn III, who noted that "In looking at the current landscape of malicious activity, the most prevalent cyber threat to date has been exploitation—the theft of information and intellectual property from government and commercial networks."[16] However, neither government nor private enterprise has fully acted on the extent of this threat.

Management in private enterprise almost always prioritizes customer experience over cybersecurity. But taking this approach can lead to irreparable damage. South Korea's largest consumer-finance firm, Hyundai Capital Services Inc., learned this lesson the hard way: Following a serious security breach, where hackers threatened to release stolen, confidential data unless a ransom was paid, the CEO now recognizes the extent of the threat and prioritizes cybersecurity; "We are now slowing down the whole organization. How things look and how they work is now secondary. Security is now first."[17]

Likewise, as a rule, government takes stronger action to help companies protect critical national infrastructure than to protect their intellectual property. The Departments of Defense and Homeland Security, for example, work together on the Defense Industrial Base (DIB) Cyber Pilot to help protect commercial suppliers to the DoD and other critical infrastructure providers from cyberattack and IP loss.[18] The Department of Energy (DoE) systematically tests the cybersecurity at power plants; a recent test in Idaho successfully breached a power plant's security and caused a generator in the plant to self-destruct.[19] The DoE also works with the nuclear industry to protect against IP theft, but we are not aware of any broader government action to protect economically important IP. The result is that more is being done to bolster cyber defenses for .mil, .gov, and critical national infrastructure than for .com.

EXHIBIT I: Cyber Threat Matrix

By combining value at stake with capability and vulnerability, we can see where the real threats are.

Estimation of value at risk relative to existing protection:
☐ Low ▨ Medium ■ High

An attack could be consistent with the attacker's incentives: •

Targets and assets

Attackers	Incentives	Public sector — Classified Information	Public sector — Sensitive Information	Public sector — Systems	Private enterprise — IP[1]	Private enterprise — Non-IP data	Private enterprise — Systems	CNI[2] — Systems	Individuals — Personal information	Individuals — Systems
Government	Strategic advantage	Med •	Med •	Med •	High •	Med •	Med •	Med • (War)	Med •	Low •
Private enterprise	Financial gain	Med •	Med •	Low	High •	Med •	Low	Med •	Med •	Low
Cyber criminals	Financial gain	Med •	Med •	Med •	Med •	Med •	Med •	Med •	Med •	Med •
Cyber terrorists, hacktivists	Protest, fun, terror	Med •	Med •	Med •	Med •	Med •	Med •	Med •	Low	Med •

1 Intellectual property
2 Critical national infrastructure, e.g., power station

The Threat to Economic Growth and Prosperity

Intellectual property and Internet-based commerce are two major drivers of U.S. economic growth and prosperity; cyberattacks threaten both. Innovative intellectual property generates significant current wealth and future growth: The World Intellectual Property Organization estimates that 45 to 75 percent of the wealth of individual companies comes from their intellectual property rights.[20] In total, intellectual property makes an estimated contribution of over $8 trillion to the U.S. economy.[21]

Similarly, the Internet is a remarkable engine for growth. A recent publication by the McKinsey Global Institute estimates that the Internet accounts for 3.4 percent of GDP in the thirteen countries examined and 21 percent of GDP growth in mature economies in the last five years. For the United States, this translates into additional total output of $440 to $580 billion, or $1,400 to $1,900 per capita—a contribution comparable to that made by the transportation, education, communication, agriculture, utilities, and mining sectors.[22]

But along with this boost to productivity and employment, the Internet brings with it new threats and vulnerabilities. Network-enabled intellectual property theft and commercial espionage threaten to undermine our national competitive advantage. Ironically, an over-energetic regulatory or bureaucratic response could be equally damaging, by constraining future web-enabled economic gains. Rather, a middle ground is needed, where government stimulates the private sector to protect its most valuable assets.

Government's Role in Protecting Private Sector Assets

Both government and business should be involved in protecting digital assets. Government needs to match its shift from a physical security mindset to a new cybersecurity mindset with a shift from "responsible for government assets only" to "responsible for key private sector assets." As Richard Clarke and Robert Knake note:

> At the beginning of the era of strategic nuclear war capability the United States deployed thousands of air defense fighter aircraft and ground based missiles to defend the population and the industrial base, not just to protect military facilities. At the beginning of the age of cyber world war the United States government is telling the population and industry to defend themselves.[23]

Government has a legitimate role in protecting intellectual property in the private sector for two reasons. Most obviously, private sector intellectual property is actually an important part of the government's own supply chain: Problems in relevant parts of the private sector are problems for the government. The best recent example of this is the alleged theft of F-35 Joint Strike Fighter data from Lockheed Martin, as the aircraft is destined for use by the U.S. armed forces and its allies. Secondly, intellectual property is an important driver of the success of the overall economy, and, as the Bipartisan Policy Centre makes clear, the success of the U.S. economy is one of the key drivers of America's global leadership: "in addition to its national security and military strength, America's global leadership derives from its economic vitality."[24] As a result, the government has a clear duty to protect. Unfortunately, the government, which has historically faced physical threats to its sovereignty and economy—threats it has countered through physical defense—is making a transition only in the areas of its sovereignty, and not in the area of its economy.

Several basic characteristics shape physical warfare. First, government can easily identify the assets it must protect (e.g., borders, bases) and the possible ways that these could be attacked. Second, the attacker or its weapons usually need to be close to the target to execute an attack. Finally, attacks are usually visible and can almost always be attributed to a specific attacker. Therefore, in physical warfare, the defender has the advantage and can put in place effective physical counter-measures.

These realities have led government to adopt a successful "perimeter approach," in which it brings key assets together and protects them behind a secure perimeter. The majority of its defenses and related investments are concentrated on fortifying the perimeter, with highest spending and newest technologies resulting in the most successful defense.

The success of the perimeter approach has, in turn, led to the development of a "physical security mindset" among decision-makers and defense practitioners. In practice, this has meant that in countering any threat (including cyberattacks), decision-makers and defense practitioners automatically default to the tried and tested physical interventions of the perimeter approach: fortify the perimeter through developing better technology and threatening retribution as a disincentive to attack.

EXHIBIT 2: PHYSICAL VERSUS CYBERSECURITY

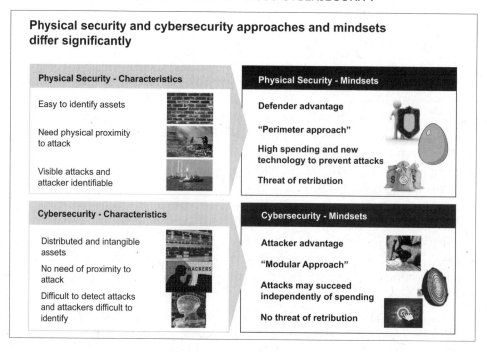

Physical security and cybersecurity approaches and mindsets differ significantly

Physical Security - Characteristics

Easy to identify assets

Need physical proximity to attack

Visible attacks and attacker identifiable

Physical Security - Mindsets

Defender advantage

"Perimeter approach"

High spending and new technology to prevent attacks

Threat of retribution

Cybersecurity - Characteristics

Distributed and intangible assets

No need of proximity to attack

Difficult to detect attacks and attackers difficult to identify

Cybersecurity - Mindsets

Attacker advantage

"Modular Approach"

Attacks may succeed independently of spending

No threat of retribution

However, the structure of the cyber environment and the threat of cyberattacks fundamentally change the rules. Focusing on perimeter security against cyberattacks is akin to building a cyber Maginot Line. These attacks are not "easily addressed by just building the security walls higher and higher"[25] because many of the characteristics of physical warfare do not apply. The defender no longer has the advantage: there is no need for proximity as the attacker can be based at any Internet-enabled computer in the world; the attacks are difficult to detect and often hard to attribute to a specific attacker; key assets to defend can be hard to identify as they are often intangible and distributed; and methods of attack are very difficult to predict. The tried and tested approaches of physical warfare (the perimeter, technological superiority, and the threat of retribution) are not nearly as effective against cyberattacks. In simple terms, technical solutions like firewalls and security software can only provide a small portion of the protection required.

In the world of physical security, the advantage falls to the defender; in the world of cybersecurity, the advantage is to the attacker. As a result, a cybersecurity mindset is required. This is characterized by an assumption that attacks will eventually breach the perimeter. This makes it important to limit the ability of an individual attack to compromise multiple assets. Doing so requires a modular approach, in which defenders divide and separate key assets, so that compromising one will not compromise the whole. A cybersecurity mindset also recognizes that attacks can come from anywhere in the world and may not be prosecutable under current laws (the U.S., for example has very little power to seek legal redress against hackers based in other countries).

To its credit, there is evidence that certain areas of government, particularly the defense and security community, are moving away from a physical security mindset. Deputy Secretary of Defense William Lynn, for example, has observed that traditional deterrence models do not apply to cyberspace.[26] Additionally, the WikiLeaks exposure did not include the government's most highly classified documents, which were held on a different system. Clearly, however, modularization could have been carried much further in this case and further limited the damage. That is to say, the transition is not complete.

Such a positive state of affairs cannot, unfortunately, be reported in the area of intellectual property, where, by and large, the government appears not to have internalized its role in helping the private sector. Given the importance of intellectual property to the continued success of the economy, this needs to change: Government and business must work together to protect intellectual property.

The Need for a "Security-Economic" Complex

Underpinning the success of the perimeter approach and hence the development of the physical security mindset was a highly successful U.S. defense industry capable of building world-leading defense solutions—what became known as the military-industrial complex. Although the perimeter approach and the military-industrial complex that supports it cannot help defend against cyberattacks for the reasons specified above, they do suggest a parallel that can defend against such attacks. We see a security-economic complex emerging that could support better cybersecurity. This new complex is a system of relationships between government, private enterprises (as the owners of intellectual property and purchasers of security solutions), and cybersecurity vendors such as Cisco, IBM, HP, McAfee, and Symantec. The security-economic complex could operate in a similar fashion to the military-industrial complex in terms of creating a set of mutually reinforcing incentives from which all parties would benefit: Government would secure economic prosperity, private enterprise would effectively protect its assets, and cybersecurity vendors would earn returns to fund future development.

In the transition to the security-economic complex, the role of government evolves from direct purchaser of defense equipment to a key stakeholder in the national economy: It still has an incentive to protect, but no longer directly purchases the solutions to do so. A fully functioning complex would help protect the economy from the threat of cyberattacks. In this state, government recognizes it has a responsibility to protect the economy, and actively seeks to help private enterprise understand the economic value at risk of the intellectual property it owns. When private enterprise understands the true extent of the threat, it will raise its level of investment in cybersecurity (in terms of both improving technical defenses and investing more in management) to counter the threat. This would increase the revenues of cybersecurity vendors and allow them to invest in developing better ways to combat the evolving cyber threat. Cybersecurity vendors also close the loop by keeping the government fully informed of the extent of the threat and lobbying it to continue its role in helping private enterprise see the full cost of the threat. Government also receives information from the security agencies that reinforce the message it is receiving from the vendors.

The security-economic complex, however, is not yet fully operational. As noted above, private enterprise and areas of government outside the defense and security community don't fully understand the extent of the cyber threat that private enterprise faces. Moreover, the government as a whole has yet to appreciate that supporting national prosperity extends to supporting the private sector in protecting its intellectual property. In practice, private enterprise underestimates how much of its value is at risk of cyberattack because neither the security agencies nor the broader government provide it with the information necessary to make this assessment.

Private enterprise is making this underestimation because it lacks two types of information, both of which could be provided by government. First, it lacks general information about the extent of the threat. Those areas of government, such as the National Security Agency, that do have good knowledge of the extent of the threat are not systematically sharing this with either private enterprise or with other areas of government. Second, private enterprise lacks specific information about vulnerabilities and specific attacks. At present, private enterprises do not know enough about the vulnerabilities in their own systems to invest sufficiently in cybersecurity. As the Sans Institute, a technology group that authors guidance for the Department of Homeland Security, put it, there is no "awareness high up in companies that there [are] such gaping holes in their software applications."[27] Private enterprises only receive this information through weak links—infrequent, ad hoc communication with the security agencies and marketing from security vendors. In addition, private enterprises perceive little, if any, incentive to share amongst themselves details of attacks they have experienced. (Although in some industries the consciousness of mutual benefit is beginning to develop, for example the Financial Services Information Sharing and Analysis Centre collects and shares cyberattack information in the financial sector; the board is formed of senior executives in the financial industry, while strategic sponsors come from the cybersecurity industry.) This second failure inhibits the development of protection approaches and technologies in a way that would not be possible with physical security, where it is visible and obvious when there has been a breach.

As a result, private enterprise underestimates its incentive to protect itself from cyberattack, which leads to misalignment of the incentives of government with those of private enterprise and vendors, and therefore insufficient protection against cyberattack. This represents a continued threat to U.S. competitiveness.

Policy Implications

Cyberattacks pose a significant threat to the continued prosperity of the U.S. economy. Government needs to play an active role in ensuring that this threat is mitigated. To do this, policymakers should drive private enterprise to protect its intellectual property adequately, and support it in doing so. This could be achieved through supporting the security-economic complex to develop it into a fully functioning system in which the economic incentives of private enterprise and cybersecurity vendors align with government's incentive to protect long-term prosperity.

To make the security-economic complex fully functional, policy changes are required in two areas. First, private intellectual property protection should be on the government agenda. Government needs to understand, and act on, its responsibility to protect both its own and private enterprise's intellectual property, as this intellectual property is vital to continued economic prosperity. To do this, policymakers need to consider how to reinforce the message within government that important assets are not only physical assets under government control, but also—and increasingly—digital, and owned by the private sector. Policymakers should also consider how they can best ensure that details held by the security agencies on the extent of the cyber threat are shared with elected officials and hence ensure that knowledge on the extent of the threat to intellectual property is well understood across government.

The U.S. would not be the first government to take action. The Australian government has authorized agencies such as the Australian Security Intelligence Organization and the Attorney General's department to work directly with private enterprise to help address information failure and threat mitigation. The Australian Defence Signals Directorate supports them in this work.[28] Something similar, possibly including knowledge transfer from government to private enterprise, could be considered in the U.S.

Similarly, government must provide incentives for private enterprise and for cybersecurity vendors. For private enterprise, these should encourage a stronger management approach in dealing with the cyber threat. For vendors, these should encourage more R&D investment to help improve technical defenses. An example might include legislation on minimum cybersecurity standards for companies.

Second, more information should be shared on the extent of the cyber threat to incentivize private enterprise to invest in management and technology to protect intellectual property. Policymakers should consider the following questions:

1. How can government and security agencies best communicate the true extent of the threat? Government must provide private enterprise with enough information on the extent and nature of the threat so that companies can understand the risks they are facing. This may include being more active

in helping private enterprise detect attacks and more readily sharing the information government has on such attacks. One possibility would be to create a venue where the government can share information with businesses on the true nature of attacks.

2. How can government best help private enterprise detect attacks? Policymakers should consider how best to develop the links between security agencies and private enterprise so that they can pass on information on actual attacks without compromising sources and methods. One possibility would be to appoint business liaison officers in security agencies to work with companies on cyber issues.

3. How can government best ensure that private enterprises report the attacks they suffer? Policymakers need to encourage the development of a framework within which companies can share details of the attacks that they have faced while minimizing any detrimental impact of such reporting on the companies themselves. An independent body could be established to anonymously collect and share details of attacks.

4. Finally, how can government make sure private enterprise puts in place managerial and technical solutions to reduce the impact of a cyberattack? Policymakers should consider establishing requirements or providing incentives to ensure that enterprises have a minimum set of cyber solutions. Here, legislation and guidelines could be effective.

The Consequences of Inaction

Technological developments are leading to an increase in attacker capabilities faster than reduction in vulnerabilities, exacerbating the attacker advantage. This asymmetry suggests that the frequency and impact of attacks will continue to increase. At the same time, the incentives to attack are growing for many attackers, not least because of the increasing amount, and value, of data available. The McKinsey Global Institute estimates that enterprises stored more than seven exabytes of new data on disk drives last year (equivalent to 28,000 times the information stored in the Library of Congress), the effective use of which is the key to productivity and margin gains.[29]

The cyber threat will not diminish of its own accord. If no action is taken, attacks will continue to increase, the value at risk will continue to grow, and U.S. competitiveness and prosperity will suffer.

John Dowdy is a Senior Partner in the London office of McKinsey & Company, where he leads McKinsey's global defense and security practice, focusing on improving the efficiency and effectiveness of defense expenditure, improving supply chain and logistics processes, and conducting stability operations in fragile states. He also leads McKinsey's joint research project, in conjunction with the London School of Economics Center for Economic Performance and Stanford Business School, on global manufacturing productivity. Prior to that, he was responsible for all of McKinsey's government work in Europe, the Middle East and Africa. He also chaired McKinsey's global efforts in economic development and served as a member of Public Services Productivity Panel. His most recent publication is *McKinsey on Government's Special Issue on Defense & Security* (spring 2010), where he contributed multiple articles including *Improving US equipment acquisition, An expert view on defense procurement* and *Stabilizing Iraq: A conversation with Paul Brinkley*. Previously, he worked as a research associate at Harvard Business School, contributing to the book *Beyond Free Trade: Firms, Government and Global Competition*. Mr. Dowdy is a fellow of the Royal United Services Institute (RUSI), a fellow of the Royal Aeronautical Society (RAeS), a member of Chatham House and a member of the International Institute for Strategic Studies. Mr. Dowdy holds a B.S. in Electrical Engineering and Computer Science from the University of California at Berkeley and an M.B.A. with high distinction from Harvard Business School.

[1] U.S. National Security Council and Homeland Security Council, *Assuring a Trusted and Resilient Information and Communications Infrastructure* (Washington, D.C., 2009).

[2] Ibid.

[3] Based on off-the-record discussions with government officials (2010-2011).

[4] As explained in Florencio Dinei and Cormac Herley, "Sex, Lies and Cybercrime Surveys," *Microsoft Research* (June 2011).

[5] Jonathan Masters, interview with Dmitri Alperovitch on the Council on Foreign Relations website, "Cybertheft and the U.S. Economy" (August 2011).

[6] Jonathan Soble, "Sony Battles Further Hacker Attacks," *Financial Times*, May 25, 2011.

[7] Sofia Mitra-Thakur, "Sony Faces Legal Costs of up to $1.5bn after Data Breach," *Engineering & Technology Magazine* (April 2011).

[8] Based on comparison of share prices on April 20 and April 25, 2011 (Sony) and May 19 and May 27 (Lockheed Martin).

[9] Based on comparison of share prices on Mar 17 and Mar 24, 2011.

[10] U.S. General Accounting Office, "Defense Acquisitions: Assessment of Selected Weapons Programmes" (Washington, D.C., 2011).

[11] InterComputer, "Massive Cyber Attack Shocks 2500 Companies," (February 2010).

[12] Dmitri Alperovitch, "Revealed: Operation Shady RAT," McAfee White Paper, (August 2011).

[13] Ibid.

[14] Stewart Baker, Natalia Filipiak, and Katrina Timlin, "In the Dark: Crucial Industries Confront Cyber attacks" a joint publication of McAfee and the Center for Strategic and International Studies (2011).

[15] Florencio Dinei and Cormac Herley, "Sex, Lies and Cybercrime Surveys," *Microsoft Research* (June 2011).

[16] Deputy Secretary of Defense William J. Lynn III, "Remarks on the Department of Defense Cyber Strategy" (Washington, D.C., July 14, 2011).

[17] Evan Ramstad, "Executive Learns From Hack: CEO Now Treats IT Department as Critical to Hyundai Capital's Operation," *Wall Street Journal*, June 21, 2011.

[18] U.S. Department of Defense, "Department of Defense Strategy for Operating in Cyberspace," (Washington, D.C., July 2011).

[19] SafteyIssues.com, "Power Plants Vulnerable to Cyber Attack" (undated).

[20] Ian Cockburn, "Assessing the Value of a Patent," *World Intellectual Property Organization* (undated).

[21] Robert Shapiro and Kevin Hassett, "What Ideas are Worth: The Value of Intellectual Capital and Intangible Assets in the American Economy," a report of the Sonecon Institute (2011).

[22] McKinsey Global Institute, "Internet Matters: The Net's Sweeping Impact on Growth, Jobs, and Prosperity" (May 2011).

[23] Richard A. Clarke and Robert Knake, *Cyber War: The Next Threat to National Security and What to Do About It* (New York: Harper Collins, 2010), p. 144.

[24] Bipartisan Policy Center, "Disciplining the Defense Budget: Lessons from a Joint BPC/Stimson Event", March 2011.

[25] UK Government Communications Headquarters Director Iain Lobban, remarks at the International Institute for Strategic Studies, London, UK, October 12, 2010.

[26] William J. Lynn III, "Defending a New Domain: The Pentagon's Cyberstrategy," *Foreign Affairs* (September/October 2010).

[27] John Gapper, "Companies Make it Easy for Hackers," *Financial Times*, June 29, 2011.

[28] Dylan Welch, "Call for More Money to Fight Growing Cyber Arms Race," *Sydney Morning Herald*, June 4, 2011.

[21] McKinsey Global Institute, "Big Data: The Next Frontier for Innovation, Competition and Productivity" (May 2011).

"In one year, five years, ten years, and twenty-five years, will we look back and see this as a time when normal market behavior and normal government actions failed to achieve our common goals? Or, will we see this as the period when goals were met through new ideas, expanded thinking, and the combined efforts of industry and government working as one?"

—MELISSA E. HATHAWAY

Falling Prey to Cybercrime:
Implications for Business and the Economy

Melissa E. Hathaway
President
Hathaway Global Strategies LLC

Espionage, "the practice of spying or using spies to obtain information about the plans and activities especially of a foreign government or a competing company"[1] is pervasive in the United States. Foreign governments and criminal networks are stealing our ideas, counterfeiting our goods, and putting our future economic well-being at risk. The number of businesses falling victim to these crimes increases daily, and no sector is without compromise. Secretary of Commerce Gary Locke recently stated that, "every year, American companies in fields as diverse as energy, technology, entertainment and pharmaceuticals lose between $200-$250 billion to counterfeiting and piracy."[2] But it is not just about counterfeiting and piracy; companies and governments regularly face attempts by others to gain unauthorized access through the Internet to their data and information technology systems by, for example, masquerading as authorized users or through the surreptitious introduction of malicious software.

We did not arrive at this place overnight. The Internet, born with its first transmission on October 29, 1969, was never conceived as the backbone of global commerce. Rather, it evolved into this role through a series of events including: (1) the first virtual data communication with Europe in 1973; (2) the first cellular portable telephone in 1973; (3) the first automated commercial cellular network in 1979; (4) the advent of the personal computer in 1981; (5) the introduction of top-level-domains (for example, .mil, .com, .edu, .gov) in 1985; (6) the creation of hyper-text mark-up language (HTML) in 1990, which enabled expanded and user-friendly information sharing on the Internet—which ultimately became the World Wide Web; (7) the relaxation of export controls for encryption products to foster global electronic commerce in 1996; (8) international adoption of the domain name system (DNS) to enable a framework for global electronic commerce; and (9) the widespread adoption of new technologies like voice over Internet protocol (1996),

WiFi (1997), wikipedia (2001), the Google search engine (1997), social networking technology (2002), and voice and video over Internet with Skype (2003). The private sector is driving innovation and adoption of technology with the promise of lower costs, increased productivity, and consumer usability without much discussion of security. In contrast, and very much the reality we face, this same technology and attendant services are being exploited for crime and conflict.

In 1988, the release and propagation of the Morris Worm affected 10 percent of the Internet's computers and disrupted Internet services for days.[3] As one of the first major "infections" experienced by both governments and businesses, it inspired the information security commodity market. Digital Equipment Corporation developed the first packet filter firewall in 1988 and so began the evolution of security products to protect us from the insecurity of doing business on the Internet.

Over the course of the next twenty years, we experienced breaches to our banks (Citibank in 1984), theft of our passwords and credit card information (AOL in 1995), penetration of the Department of Defense unclassified networks (Solar Sunrise in 1998), theft of our personal identifiable information (Choice Point in 2005), illegal copying of defense industrial base critical program information (weapon system designs in 2007 and ongoing), penetration of the Department of Defense classified networks (Buckshot Yankee in 2008), and targeting of our children (Sony 2011). These and other "cyberattacks on Internet commerce, vital business sectors, and government agencies have grown exponentially. Some estimates suggest that in the first quarter of 2011 security experts were seeing almost 67,000 new malware threats on the Internet every day. This means more than forty-five new viruses, worms, spyware, and other threats were being created every minute—more than double the number in January 2009. As these threats grow, security policy, technology, and procedures need to evolve even faster to stay ahead of the threats."[4] A recent Symantec report indicates that these trends will continue.[5] From 2010 to 2011 the differences are discouraging. In fact: There were 286 million unique variants of malware that exposed and potentially exfiltrated our personal, confidential, and proprietary data; each data breach exposed, on average, 260,000 identities; there was a 93 percent increase in web-based attacks (compromised/hijacked websites that would infect individuals' computers if visited); the underground economy paid anywhere from $.07 to $100 for our stolen credit card numbers; and realizing that mobile payments and mobile platforms (like smart phones and the iPad) would be the newest vector of technology adoption, there was a 42 percent increase in mobile operating system vulnerabilities and subsequent exploitation.

As American businesses, inventors, and artists market, sell, and distribute their products worldwide via the Internet, the threat from criminals and criminal organizations who want to profit illegally from their hard work grows. The threat from other nations wanting to jump start their industries without making the intellectual investment is even more disturbing. This fleecing of America must stop. We can no longer afford complacency and silence—we must find and use as many market levers as possible to change the path we are on.

This chapter discusses three different approaches to addressing the problem. First, it is possible to apply tax incentives to businesses for innovation and consumers to patch the problem. While this may not be a fiscally responsible approach given the current debt crisis and constrained economic environment, it should nonetheless be considered. Second, Congress could consider reviewing the applicability of the National Defense Production Act to provide our IT industry a fighting chance against the predatory pricing and industrial espionage being practiced by other nations. Finally, this chapter will discuss four unique private-public partnerships that deserve attention as regional and potentially national agents of change.

Use Tax Incentives

It is estimated that the G-20 economies have lost 2.5 million jobs to counterfeiting and piracy, and that governments and consumers lose $125 billion annually, including losses in tax revenue.[6] The underground economy makes it easy for anyone to get started in cybercrime. The tools and services are readily available to take advantage of the average consumer and exploit the industry's latest product. So why can't we help our information security industry innovate as fast as the criminals? The research and experimentation credit under section 41 of the Internal Revenue Code provides a tax credit for incremental investment in research, which could be applied to address this innovation gap. There is a 20 percent credit for incremental research over a base period, or an alternative simplified credit of 14 percent for incremental research over the previous three years. Originally enacted in 1986, the research credit is a temporary provision that must be extended regularly. In its FY2012 budget proposal, the Obama administration proposed to make the research credit permanent, and to increase the credit percentage on the alternative simplified credit to 17 percent.[7] The research credit is not specific to any particular type of research or industry, but is available for any research that is technological in nature. So, just as the Digital Equipment Corporation introduced some of the first technology (the firewall) to address exploitation of our systems, our industry could apply its research toward monetizing products that begin to close the gap between criminal exploitation and successful protection. The innovation agenda could also be applied to data correlation, detection of network and system anomalies, and identification and evidence gathering of criminal "fingerprints."

Because the research credit is focused on basic research, it serves as an incentive for companies to develop new ideas that can be deployed in their business. However, the credit does not extend to purchases of products or technologies that are used in a business. If an incentive is needed to encourage companies to acquire tools that can be used to enhance their cybersecurity, the research credit will not suffice. Rather, Congress could consider tax incentives to encourage taxpayers to acquire and deploy new security tools by providing an investment credit to encourage such investments. For example, since 1992, Congress has provided incentives for taxpayers to invest in renewable energy through the energy investment credit under section 48 of the Internal Revenue Code and through the renewable energy production credit under section 45. These incentives have helped to encourage taxpayers to deploy resources to develop wind, solar, geothermal, and other types of renewable energy. Similar programs could be implemented to assist in the development of new tools to protect the security of our information and communications infrastructure.

While tax credits can help incentivize taxpayers to focus investment on favored items, a tax credit is only useful to a taxpayer with positive taxable income who can use the credit to shelter the tax burden on that income. For companies in depressed markets, with net operating losses, a tax credit has no value and will not provide any incentive for new investment. In recognition of this problem, in 2009 Congress provided a temporary program for grants in lieu of the low-income housing and energy credits.[8] A similar type of temporary grant program might be appropriate to kick start intensive investment in technology to improve cybersecurity.

Whether a credit or a grant program is established, key drafting considerations would need to ensure that the benefits are provided only for new investment in the type of technology that Congress wants to incentivize, that the investment is made in the United States, and that research and manufacturing relating to the favored products is conducted in the United States. In addition, to ensure that the benefits are used solely for innovation, the provisions should be drafted to ensure that only the taxpayers who invest in and deploy the technology can receive the benefits provided. This would stand in contrast to the low-income housing and energy credit programs that have been developed over the years to permit financial investors to take advantage of their benefits.

Offering a similar incentive to the average consumer may also go a long way toward improving the nation's security posture. Because consumers may not keep pace with the latest technology improvements or band-aids in security, Congress also should consider providing targeted incentives for consumers to invest in securing their personal computers and home networks. Again, the energy sector provides a useful precedent: Section 25C of the Internal Revenue Code provides a credit of up to $500 per year for individual investments in residential energy efficiency improvements. This credit has encouraged investments in energy-efficient appliances, HVAC systems, and windows and doors. Separately, section 25D provides a 30 percent investment credit for investments in residential solar, wind, and geothermal systems. And over the past decade the hybrid vehicle industry has flourished in the United States, in large part due to the tax credits provided to incentivize these purchases. In the case of cyber investments in the home, the average dollar cost per household is relatively small, so tax credits may not impact the economic decision as much as in the case of the energy examples described above. Nevertheless, a credit of even $25 per taxpayer who purchases new security software each year could help further proliferate these important safeguards.

Leverage the Authorities in the National Defense Production Act (NDPA)

In addition to using taxes as a market incentive, Congress should also consider applying the NDPA to counter the broad based espionage being conducted against our defense industrial base coupled with the predatory pricing and acquisition strategies of our core telecommunications technologies by foreign corporations. Foreign companies are gaining an ever-increasing share of the U.S. commercial technology market, while at the same time our national security networks, critical infrastructure, and weapons systems are growing more reliant on products and services from that market. This is further complicated by the fact that China is our largest supplier of telecommunications imports (42 percent) and is our eighth largest export market for U.S.-based telecommunications technologies. The NDPA could be applied in the absence of industrial policy or market levers that can shore-up the competitive position of U.S.-based information and communications technology (ICT) companies.

In response to the start of the Korean War, the NDPA was enacted in 1950 as part of a broad civil defense and war mobilization effort in the context of the Cold War. The act contained seven sections, of which three major sections remain active today. The first (Title I: Priorities and Allocations) authorizes the president to require businesses to sign contracts or fulfill orders deemed necessary for national defense. The second (Title III: Expansion of Productive Capacity and Supply) authorizes the president to establish mechanisms (such as regulations, orders, and agencies), to develop, modernize, and expand defense productive capacity. The third area (Title VII: General Provisions) provides antitrust protection for voluntary industry agreements serving defense interests, and established a voluntary reserve of trained private sector executives available for emergency federal employment.[9] Beginning in the 1980s, the Department of Defense (DoD) began using the contracting and spending provisions of the NDPA to provide seed money to develop new technologies. Using the NDPA, DoD assisted in the development of a number of new technologies and materials, including silicon carbide ceramics, indium phosphide and gallium arsenide semiconductors, microwave power tubes, radiation-hardened microelectronics, superconducting wire, and metal composites.

In the late 1980s and early 1990s, U.S. industry faced fierce competition in the area of micro-electronics, specifically with semiconductors from Japan. The U.S. government co-invested with industry to establish Sematech to upgrade the production environment and improve quality and yield of product to market. New technologies create new opportunities and one could argue these investments led to many of the micro-electronics that are part of the American household today, including cell phones, netbooks, and iPads, among others.

The information technology industry is critical to the economic and national security of the nation, much as the aerospace industry was crucial to our security posture during the 1960s. The pace of innovation and marketplace dynamics are threatening U.S. leadership in communications, computing, networking, and security technologies, and it may be time to provide government assistance to enhance the competitiveness and preserve the leadership of this critical sector. For example, Congress could leverage the special authorities contained in the NDPA to help subsidize and accelerate DoD access to commercial production technologies and capacity. The NDPA also provides for anti-trust protection for voluntary agreements among business competitors to enable cooperation to plan and coordinate measures to increase the supply of materials and services needed for national economic and defense purposes. The NDPA also authorizes the establishment of the National Defense Executive Reserve (NDER) (what some would call the Civilian Cyber Reserve Corps), a cadre of persons with recognized expertise who could step into executive positions in the Federal government in the event of an emergency. One could argue that the Department of Defense information technology exchange program (ITEP) initiative could be the long-term pipeline for this NDER. The government should recognize that our national telecommunications infrastructure is vital to U.S. interests and consider better protecting it. Any discussion of government protection of the industry should include the primary and subsidiary providers and suppliers. Furthermore, the government should consider a broad definition of the IT environment, to include current and future converged communications, infrastructures, and services. It may be wise to draw upon the Electronic Communications and Privacy Act (ECPA) definition: "including voice over Internet-Protocol communications; by the aid of wire, cable, or other like connection including wireless connections such as mobile phones, satellites, and fiber-optic cables."[10]

It is desirable to use Title III authorities to upgrade suppliers' production capabilities to improve quality and yield on new technologies that would enhance the security of our critical infrastructures, networks, and mobile devices while at the same time making our IT corporations more competitive. Example areas for technology investments include: systems architectures that permit the secure use of commercial-off-the-shelf (COTS) computers, software, and networks; mechanisms, including intelligent agents, for locating and retrieving information from complex database structures; automated systems for reverse engineering based on scanning of an actual part; design of interruption-free connector systems for ultra-high-speed data rates; high performance computing (HPC) and advanced visualization of petabyte data sets; advanced visualization; and environments to perform at scale network simulations and rapid prototype testing.[11]

Why should we explore the NDPA option? The United States' ability to project power is wholly reliant on the strength of our IT sector. Other countries (for example, China and Russia), recognizing the importance of the IT industry to their overall national economic health, are pursuing strategies that support their IT industry leadership. The United States needs to find equivalent market levers to shore up our indigenous IT companies and help drive focused research and development (R&D) for the next generations of innovation with the goal of building a more secure, resilient infrastructure.

Accelerate and Seed Private-Private and Private-Public Initiatives

Finally, as discussed above, the proposed tax incentives coupled with the NDPA could enhance emerging private sector initiated partnerships and innovation to close the security gap. These grassroots efforts are being initiated by businesses who can no longer tolerate being victimized by criminals and foreign governments alike. Each program aspires to reduce the overall incidence and harm caused by cyber incidents and each program is improving collaboration and operational information-sharing while simultaneously protecting sensitive data and ensuring the security of the broader community. Four initiatives—the Cyber Accelerator, the Network Security Innovation Center, the Advanced Cybersecurity Center, and the National Economic Security Grid—are discussed in detail below.

The Cyber Accelerator is a structured consortium that uses DoD's transaction authority to invert the acquisition model from pull- to push-sourcing and repurposes private sector innovation to meet DoD's needs. The government enters into a technology investment agreement (TIA) with the non-profit consortium lead to assist in research and development of commercial technologies to apply to DoD use cases and defense technology allowing tech transfer of intellectual property to commercial entities. The goal of the effort is to expand integration of innovative technologies within the commercial marketplace (for example Google, Intel, McAfee, and VMware) to add value beyond the large-scale system (weapon system) integrators (Lockheed Martin, General Dynamics, and Northrop Grumman). The Cyber Accelerator seeks to lower the private sector barrier to working with the DoD while simultaneously providing the DoD with shorter product cycles, lower life cycle costs, and privileged access to commercial innovation.

Two benefits converge to open up a new set of vendors and new innovation for the government. First, identifying and funding the development of "dual-use" capabilities attracts private sector investment and at the same time addresses an operational and technical shortfall in DoD. Second, it attracts companies that are reticent to deal with DoD by protecting their intellectual property and seeding capability development that leads to both company and investor profits and DoD operational needs.

The work of the consortium follows an agreed upon multi-year technology roadmap with an annual funding plan that complies with authorities and appropriations. Some technology initiatives that this effort will explore include: (1) enhanced authentication (endpoint, application, and data); (2) identity and behavior recognition (correlating user behavior across multiple personas, devices, and accounts); (3) trusted data provenance (tied to identity for source, application, and user roles); and (4) automated learning for remediation and response.

Lessons learned from these dual-use product initiatives will provide the government with insight for future acquisition reform and potential innovative models. It also provides a mechanism to use market incentives for rapid innovation and deployment.

The Network Security Innovation Center (NSIC) is an industry driven initiative based out of Silicon Valley to create a government, academic, and industry partnership to foster innovation and information-sharing in cybersecurity. The NSIC brings together the talent of the largest IT companies and entrepreneurs of the Bay Area with the computational capacity and unique capabilities of the FFRDC (Federally Funded Research and Development Center) status of Lawrence Livermore National Lab (LLNL). This initiative is striving for "extreme security innovation," says Jacques Francoeur, executive director of the Bay Area CSO Council, who played an instrumental role in bringing industry stakeholders to the table, using intellectual power, computational power, and most importantly industry power.[12] In a recent speech at the center, Gary Terrell, Adobe's chief information security officer, described the top strategic initiatives businesses must launch to meet the growing threats of worldwide cybercrime, stating that, "security leadership needs to fundamentally change its perspective and, in many cases, make a 180-degree turn to protect their digital assets, and the time is now." [13]

The NSIC has two "anchor" IT firms initiating focused collaborative R&D projects. These projects are indicative of what the center could offer, as an incubator and a direct path for moving R&D results to sustainable innovative products. For example, McAfee, which has an Internet threat sensor network collecting data in 120 countries, and LLNL are jointly working on probability models for dynamics on graphs. They are trying to run analytics to winnow out critical threats from this massive data set (of 100 billion queries per month), to see if they can find interesting patterns with significantly greater computing capacity. By partnering with LLNL, McAfee gains access to the lab's supercomputing and highly specialized scientific resources, allowing it to handle large data analysis requirements and potentially enabling McAfee to develop new technologies to counter advanced threat techniques, profile hackers, and insider threats. Cisco is also partnering with LLNL on a focused project regarding network simulation and virtualization. The goal of this project is to simulate large-scale exploits without disrupting operational networks in order to discover the second-order effects of exploits with the aim of developing techniques for early detection. By having a large-scale network simulator that has access to large volumes of real world data (provided by the Cisco IOS platform that is currently operating on millions of active systems, ranging from the small home office router to the core systems of the world's largest service provider networks), it may be possible to create an environment and technology that can lead to attacker attribution. The simulation creates flexible honeypots and "hacker treadmills" to keep adversaries engaged while allowing the time and interactions required to gain attribution.

The NSIC is working to become fully operational in the next six months. It is currently in the process of defining a governance framework and intellectual property rights model that meets the needs of all parties. The NSIC shows great promise as an innovation engine that addresses some of our toughest cybersecurity problems, especially around big-data and malicious behavior analysis.

The Advanced Cybersecurity Center (ACSC) was created to establish Massachusetts—and the New England region more broadly—as a leader in the development of next generation cybersecurity R&D and education programs. This industry-driven initiative brings together university and government entities to address advanced cyber threats by sharing insights on attacks and mitigation strategies and cultivating the next generation of talent for employment in the region. The ACSC supports a collaborative, cross-sector research environment (and facility) using the region's unparalleled university, research, and industrial resources to focus on areas not addressed by commercial security solutions and thereby strengthen members' defensive capabilities. The ACSC has formed working groups to drive the Center's collaborations across a range of initiatives including: (1) threat evaluation and data sharing, (2) university-industry partnerships, and (3) policy and legal challenges. Specific technology projects seek to enable trusted collaborations in the pre-competitive space and foster innovations and improvements in predictive analytics, incident monitoring and analysis, intrusion detection and eradication, and deployment, incident scenarios and response strategies.

As the ACSC becomes operational it intends to establish federal and national partnerships to extend the region's influence and enhance coordination with key resources, becoming a vital component in protecting the region's and nation's key assets.

Finally, the National Economic Security Grid (NESG)[14] is a grassroots-based independent non-profit organization established in 2010 as a resource for metropolitan area public and private sector entities. The NESG is committed to dedicating resources and capability to local small and medium enterprises (SMEs) in each of the metropolitan areas across the country and providing them with information, processes, proven practices, and solutions to the risk, threats, and hazards they face every day. The goal is to establish NESG operations in every metropolitan area in every state across the country to truly create a "National Economic Security Grid."

NESG selected metropolitan Los Angeles as its inaugural site after LA County Sheriff Leroy Baca expressed a strong desire to launch this grassroots initiative as a means of strengthening the local partnership between the public and private sectors, with a focus on safeguarding the economic security of the city. As such, the NESG will collect "intelligence" data on a broad range of external and internal risks, threats, and hazards that may affect the local SME community and will turn this data into tailored actionable information for delivery to online secure escrow accounts accessible by SME members. It also plans to establish a Risk Solution Center that provides tested and vetted risk mitigation solutions to SMEs.

While not yet fully operational, the NESG intends to make a difference by: (1) establishing strong local partnerships between SMEs, local law enforcement, prosecutors, politicians, and other community-based support groups to focus on the stability, viability, and resiliency of the local community and its economic environment; (2) providing actionable information to SMEs on the real world risks and threats they face every day; and (3) identifying sound and affordable risk mitigation solutions to ensure high survivability of SMEs, which ultimately improves the economic conditions of the community.

Conclusion

Notwithstanding all of the efforts made to date by many well-intended professionals and organizations, and despite significant advances in technology, we are still struggling to stay on top of the cybersecurity problem. Indeed, the problem is growing faster than the solution and we cannot afford to be faced with strategic surprise as we falter in addressing it. The national economic security agenda for the United States needs disruptive ideas that reinvigorate our innovation engine, our intellectual creativity, and our law enforcement capability and capacity.

We need to expand our options, and to do so quickly. In one year, five years, ten years, and twenty-five years, will we look back and see this as a time when normal market behavior and normal government actions failed to achieve our common goals? Or, will we see this as the period when goals were met through new ideas, expanded thinking, and the combined efforts of industry and government working as one?

We cannot continue along the current path and expect to make adequate progress to confront the cybersecurity dilemma. Our country has at its disposal market levers, unique authorities, advanced technology, public-private partnerships, and a culture of innovation and creativity. The full gambit of market levers—especially incentives-based levers—is needed to advance research and development, drive innovation, and close the gap between adversary successes and industry defenses. We need a more secure resilient infrastructure. Can we find the wherewithal to stop the fleecing of America?

Melissa Hathaway is President of Hathaway Global Strategies LLC and a Senior Advisor at Harvard Kennedy School's Belfer Center. Ms. Hathaway served in the Obama administration as Acting Senior Director for Cyberspace at the National Security Council and led the Cyberspace Policy Review. During the last two years of the administration of George W. Bush, Ms. Hathaway served as Cyber Coordination Executive and Director of the Joint Interagency Cyber Task Force in the Office of the Director of National Intelligence where she led the development of the Comprehensive National Cybersecurity Initiative (CNCI). At the conclusion of her government service she received the National Intelligence Reform Medal in recognition of her achievements. Previously, Ms. Hathaway was a Principal with Booz Allen & Hamilton, Inc., where she led two primary business units: information operations and long range strategy and policy support, supporting key offices within the Department of Defense and intelligence community. Earlier in her career she worked with Evidence Based Research, Inc. and the American Foreign Service Association. Ms. Hathaway is a frequent keynote speaker on cybersecurity matters, and regularly publishes papers and commentary in this field.

[1] *Merriam-Webster's Dictionary*, 11th ed..

[2] Secretary of Commerce Gary Locke (Remarks at the Washington International Trade Association, Washington, D.C., July 22, 2009).

[3] Eugene H. Spafford, "The Internet Worm Program: An Analysis," *Purdue Technical Report* CSD-TR-823 (West Lafayette, IN: Purdue University, 1988).

[4] United States Department of Commerce, Internet Policy Task Force, Cybersecurity Green Paper (Washington, D.C.: Government Printing Office, June 2011), ii.

[5] Symantec Internet Security Threat Report, *Trends for 2010*, (April 2011).

[6] Frontier Economics, *Estimating the Global Economic and Social Impacts of Counterfeiting and Piracy* (February 2011).

[7] Joint Committee on Taxation, *Description of Revenue Provisions Contained in the President's Fiscal Year 2012 Budget Proposal (JCS-3-2011)* (June 2011).

[8] Sections 1602 and 1603 of the American Recovery and Reinvestment Act of 2009.

[9] Congressional Research Service, *Defense Production Act: Purpose and Scope* (May 14, 2009).

[10] 18 U.S.C. § 2510(1) (2006) and Nicholas Matlach, "Who Let the Katz Out? How the ECPA and the SCA Fail to Apply to Modern Digital Communications and How Returning to the Principles in Katz v. United States Will Fix It, 449 *Commlaw Conspectus*, Vol. 18 No. 2 (2010), 443.

[11] National Research Council, *Defense Manufacturing in 2010 and Beyond: Meeting the Changing Needs of National Defense* (Washington, D.C.: National Academies Press, 1999), 45.

[12] Author interview with Jacques Francoeur, July 28, 2011.

[13] Quoted in Brian D. Johnson, "Network security innovation center kicks off lecture series," on the Lawrence Livermore National Lab website, July 18, 2011.

[14] Information presented here is derived from discussions with the founder, Lynn Mattice.

Part 4

CYBERSECURITY AND ITS TENSIONS WITH INTERNET FREEDOM

CHAPTER 7

Internet Freedom and its Tensions with Cybersecurity

Richard A. Falkenrath
Principal
The Chertoff Group

CHAPTER 8

Internet Freedom and Political Change

Richard Fontaine
Senior Advisor
Center for a New American Security

"*This tension [between Internet freedom and the security of cyberspace] is a technological manifestation of America's long-standing ambivalence between idealism and realism, compounded by the extraordinarily high economic stakes of Internet governance and complicated by a legal system that has not and indeed cannot keep pace with rapid technological change.*"

—RICHARD A. FALKENRATH

Internet Freedom and its Tensions with Cybersecurity

Richard A. Falkenrath
Principal
The Chertoff Group

The U.S. government's support for Internet freedom is nearly absolute, at least rhetorically. According to Secretary of State Hillary Clinton, speaking in January 2010, "we stand for a single Internet where all of humanity has equal access to knowledge and ideas."[1]

President Obama, during his November 2009 visit to China, was a little more equivocal. "The more freely information flows," he said, "the stronger the society becomes, because then citizens of countries around the world can hold their own governments accountable… [T]here's some price that you pay for openness, there's no denying that. But I think that the good outweighs the bad so much that it's better to maintain that openness."[2]

Cybersecurity is also important to the U.S. government. "From now on," President Obama declared in May 2009, "our digital infrastructure—the networks and computers we depend on every day—will be treated as they should be: as a strategic national asset. Protecting this infrastructure will be a national security priority. We will ensure that these networks are secure, trustworthy, and resilient. We will deter, prevent, detect, and defend against attacks and recover quickly from any disruptions or damage."[3]

These two priorities—Internet freedom[4] and the security of cyberspace—are often in tension. This tension is a technological manifestation of America's long-standing ambivalence between idealism and realism, compounded by the extraordinarily high economic stakes of Internet governance and complicated by a legal system that has not and indeed cannot keep pace with rapid technological change.

Understanding the instruments of Internet repression

So what are the instruments used by repressive governments to suppress Internet freedom? They are many, they are diverse, they are opaque, and they are evolving rapidly to keep pace with technological change and changes in the way people interact in cyberspace.

Traditional enforcement (Censorship 1.0)

The simplest means of government repression is for government agents to locate individuals producing, transmitting, storing, or consuming objectionable content and to punish them. The techniques of traditional enforcement are, of course, employed to a wide range of ends, not just political repression. Their efficacy depends on the government's ability to reach the individuals in question. In certain respects, the Internet allows the producers and disseminators of objectionable content to reside beyond the reach of government and in some cases to remain anonymous; in other cases, certain applications on the Internet—including many social media services, especially those with real-name registration and geolocational features—make it easier for the government to identify and locate people based on their online activities.[5]

Electronic surveillance (Censorship 1.1)

Repressive governments almost always employ electronic surveillance of some form or another, in concert with traditional enforcement, as part of their overall censorship regime. And even when the authorities are not monitoring Internet traffic, the expectation that they are doing so—or that they easily could—can have a chilling effect on the expression of dissent.

Electronic surveillance on the Internet is an immensely complex subject, but three broad issues are particularly important for the present discussion:

- *Anonymity.* The government's ability to quickly and accurately relate online identity to real-world identity is the chief determinant of its ability to take enforcement action against suspected perpetrators of illegal activity. A pro-Internet freedom policy will generally seek to make it easier for individuals in cyberspace to shield their real-world identities by, for instance, not requiring real-name registration and authentication for Internet services.

- *Retention periods.* In general, regardless of the procedures and criteria that control electronic surveillance by the government, the longer internet service providers (ISPs) and online application providers maintain business records and content, the better it is for governments seeking to investigate Internet activity, and the worse it is for individuals who are doing, writing, or reading things the government opposes. The importance of retention periods rises as more and more data moves off of local storage and into various cloud services.

- *"Backdoors."* Finally, when governments engage in electronic surveillance, they want easy access into the communications systems and services used by their targets. They often require such "backdoors" as a condition of regulatory approval. Novel, proprietary Internet-based communications systems—which frequently offer only an online portal without any physical in-country infrastructure—present governments with particularly difficult surveillance challenges, as do services that host only strongly encrypted data while users maintain sole possession of the necessary decryption keys. A pro-Internet freedom policy will stand against "backdoor" requirements.

One of the dilemmas of promoting Internet freedom as part of U.S. foreign policy is that, on each of these issues, the practices of the United States sit uncomfortably with a pro-Internet freedom agenda. For instance, in response to the release of over 240,000 classified diplomatic cables on WikiLeaks, the Department of Justice compelled Twitter via court order to divulge the real-world identities of individuals involved in the dissemination of these documents.[6] Similarly, the Obama administration is currently seeking legislation that would require U.S. ISPs to maintain certain categories of business records for up to two years.[7] And a 1994 federal law, the Communications Assistance to Law Enforcement Act (CALEA), requires telecommunications providers and manufacturers—including, since 2003, Internet broadband and voice carriers—to incorporate the technical means for lawful electronic surveillance by government agencies.[8] The differences between the United States and repressive governments in the area of electronic surveillance turn on the purpose of its use and the legal processes, restrictions, and protections associated with it, not on the techniques or technologies themselves.

Bandwidth throttling (Censorship 1.2)

One particularly crude technique used by governments seeking to control their public's access to information is to restrict the bandwidth available to or from local ISPs. This practice, known as "throttling," does not eliminate the public's Internet access, but can slow it to the point of being virtually useless, especially when accessing larger files, such as images, audio, and video. Throttling has been employed intermittently by Iran, in part in recognition of the limited efficacy of Iran's other censorship strategies.[9] Throttling is also, however, employed by U.S. ISPs against subscribers suspected of illegally downloading copyrighted material as part of an arrangement with the Recording Industry Association of America (RIAA) and the Motion Picture Association of America (MPAA).[10]

Filter-and-block (Censorship 1.3, 1.4, etc.)

The most prevalent method of direct Internet censorship is to filter Internet traffic and block transmissions that meet predetermined criteria. There are many different ways of implementing a filter-and-block strategy.[11] These include blocking certain IP addresses at the ISP or international gateway level; blocking pre-determined uniform resource locators (URLs) at the ISP or domain name system (DNS) level; or scanning URLs for specific strings of text or keywords, and blocking those that meet the criteria. These approaches depend on having an up-to-date list of IP addresses, URLs, or keywords that contain content deemed objectionable by the censor. Because they target only the Internet resource locational information, not the content of the transmission,[12] these approaches are relatively easy to implement, but can usually be circumvented; except when applied in an extremely broad manner (for example, a blanket ban on YouTube or all sites from or to a specific top-level domain).

The filter-and-block systems employed by governments to control the distribution of objectionable information inside their borders are similar and in some cases identical to the systems employed by libraries and schools to limit access to pornography and by U.S. ISPs to limit illegal access to copyrighted material. The hardware, software, and services needed for a nationwide filter-and-block system are sold by a wide variety of companies; the U.S. government does not control the export of this technology.[13] Thus, when the United States objects to another country's filtering and blocking of the Internet, it is not really disputing the method itself, but the purpose to which it is applied.

Censorship through intermediaries (Censorship 2.0)[14]

Increasingly, governments are turning to the private-sector providers of online information infrastructure or services to control the content their publics consume and produce. This trend is driven by a variety of factors, the most important being the hopelessness of doing it any other way. This is especially true in the world of social media and cloud computing, where any individual user can post content of virtually any kind—including SMS files, emails, pictures, audio files, video files, geolocational tags, proprietary-format files (e.g., interactive video game avatars), encrypted files, and peer-to-peer data sharing files (e.g., torrents)—to an Internet venue that can be accessed by any other individual Internet user, anywhere in the world. The only place where data of this kind can possibly be monitored is on the servers of the company or organization providing the service, which may or may not be inside the country whose government wishes to control the information.

Censorship through intermediaries raises many novel and difficult questions, placing a number of the major information technology companies, many of them incorporated in the United States, in the position of having to make decisions that resemble U.S. foreign policy and impact U.S. national interest, but without any of the traditional U.S. policymaking inputs and constraints.

Google's experience in China is perhaps the most famous example of this phenomenon but is by no means unique. In 2006, Google reached an agreement with the Chinese government allowing it to create google.cn, a search engine built for the Chinese market, the largest in the world by number of users. As a condition of entering this market, Google agreed to censor certain content deemed objectionable by the Chinese authorities from its search results. The company argued that the net effect would be to make more information available to more people. Over the ensuing years, both sides grew progressively more uncomfortable with the arrangement, with Beijing increasing its censorship demands of Google and Google resisting. In January 2010, prompted by the revelation of a breach of Google's confidential data that was traced back to China, Google announced it would route all searches originated by mainland users through servers in Hong Kong, ostensibly out of the reach of mainland Chinese censorship laws. Later in 2010, the dispute was contained when Beijing renewed Google's Internet provider license without additional conditions for censorship cooperation, giving Google a more agreeable environment in which to resume Google.cn operations.

There is no shortage of examples of states dealing with objectionable content on the Internet through private-sector intermediaries. Sometimes it involves an explicit, direct, and transparent instruction clearly rooted in applicable local law and articulable standards, as when the governments of Germany and France direct search engines in their markets not to return results related to Nazi propaganda. Other times it results from informal consultations or understandings; perhaps of a coercive nature, perhaps not. Consider the role of private-sector intermediaries after WikiLeaks released its cache of classified U.S. diplomatic cables: The U.S.-based DNS company responsible for translating .org URLs into numeric IP addresses dropped wikileaks.org from its service, making it more difficult for people to reach the site; and a number of U.S.-based financial services companies, including PayPal, Mastercard, and Visa, stopped processing donations and payments to bank accounts connected to the WikiLeaks organization.[15] The companies explained these actions with reference to their pre-existing subscriber agreements, and there is no evidence that these steps were directed or even requested by the U.S. government. However, by the time these steps were taken, the U.S. government had made clear its position that, as stated by State Department Legal Advisor Harold Koh, the possession of classified documents by WikiLeaks represented an ongoing violation of U.S. law and "endangered the lives of countless individuals."[16] It seems plausible that the position of the U.S. government had some bearing on the decision-making of WikiLeaks' DNS and financial intermediaries.

Google's experience in China illustrates many of the new problems that arise from reliance of governments around the world on non-governmental intermediaries—often large, U.S.-based technology companies—to accomplish their objectives online. Five stand out. First, the intermediaries confront truly vexing and complex commercial, ethical, legal, and technological dilemmas.[17] Second, governments have the upper hand. Despite the aspirations of cyber-libertarians for a borderless Internet,[18] governments worldwide have successfully asserted control over the ability of legitimate companies to provide content and services over the Internet inside their borders.[19] Compliance with a government's onerous, sometimes distasteful requirements can be, and often is, the price a technology company must pay to do business in that country. According to a classified U.S. government cable revealed by WikiLeaks, "in the past, a lot of [Chinese] officials worried that the Web could not be controlled... But through the Google incident and other increased controls and surveillance, like real-name registration, they [the Chinese government] reached a conclusion: the Web is fundamentally controllable."[20]

Third, as in all economic sectors, when a government imposes requirements on a foreign company operating in its market, homegrown competitors may be inadvertent or intended beneficiaries. Some of Google's lost market share in China seems to have gone to Baidu, a search engine based in China.[21]

Fourth, interactions of this sort are carried out almost entirely behind the scenes. Neither the governments nor the intermediaries, particularly if they are private companies, have strong incentives to be transparent about the conversations they are having or the agreements they reach. The Google experience in China is unusual because much—but by no means all—of it became public.

Finally, large U.S.-based technology companies confront no shortage of domestic pressures, demands, regulatory requirements, expectations, and risks that are themselves often inconsistent and contradictory. For this reason, these companies are often hesitant to ask the U.S. government for advice or assistance. Google reportedly asked the U.S. government for help with Beijing, but Washington had little to say on the matter until after the massive Chinese-origin hacking of Google's servers was publicly disclosed in January 2010.

When U.S.-based technology companies like Google have trouble with sovereign governments like China, U.S. national interests are often so confused and conflicted that the upsides to the U.S. government taking any sort of position on the dispute rarely outweigh the downsides. Even when Washington does intervene, it quickly confronts the limitations of its own influence as well as the contradictions between American foreign policy preferences and domestic practices.

Packet inspection (Censorship 3.0)

The next frontier of electronic surveillance, cybersecurity, and copyright enforcement is something called packet inspection. Unfortunately, it is also the next frontier of censorship.

The Internet is a packet-switched network. In simplest terms, an electronic file is disassembled into many, much smaller units, which are sent to their destination through a network of switches and routers and are ultimately reassembled into the original, coherent electronic file according to one or more protocols. This architecture makes the Internet extremely difficult to disrupt—which was, of course, the consideration that drove its development during the Cold War—but also makes Internet transmissions difficult to monitor at the content level. Hence, the filter-and-block censorship strategies described above, which target the various headers and directional information attached to individual packets that are needed to move them through the network, can be relatively easily circumvented.[22]

Packet inspection is an emerging technology that scans Internet content as it is being transmitted—but before it is reassembled at a proxy server or its destination—for code or data that meet pre-determined criteria in order to facilitate electronic surveillance or block malicious or objectionable content. Like censorship through intermediaries, and for obvious anti-circumvention reasons, the network operators that are deploying packet inspection technology are not transparent about its capability, screening criteria, or extent of utilization. Nevertheless, it is known that:

- Packet inspection is an element of the more advanced cybersecurity strategies being developed by the U.S. government under the rubric of "active network defense;"[23]

- It is being examined, and possibly used, in concert with U.S. ISPs, by the RIAA and MPAA as a method for enforcing copyrights;[24] and

- U.S. companies are developing, marketing, and exporting these technologies.[25]

Like filter-and-block censorship strategies, the efficacy of packet inspection technology depends in part on the ability to define criteria to screen against. This will not always be possible; which, combined with the enormous volume of information transmitted through the Internet, means that packet inspection technology will not suddenly make the Internet dramatically easier to monitor. It does, however, indicate the technological direction in which Internet censorship, cybersecurity, electronic surveillance, and intellectual property enforcement is moving.

Implications for U.S. policy

As the preceding discussion hopefully makes clear, the instruments used by repressive governments to suppress Internet freedom are similar to—and in some cases identical to—those used or allowed by democratic governments to achieve a variety of lawful or socially acceptable ends, including securing critical computer networks. The immense importance currently being attached to cybersecurity in the United States accentuates these contradictions, since it has accelerated the adoption by U.S. national security agencies of the technological and operational capabilities used as instruments of political repression elsewhere in the world. This fact has a number of important U.S. policy implications.

First, when policymakers consider how to incorporate Internet freedom into U.S. foreign relations, they would do well to bear in mind just how unsettled, confused, and conflicted U.S. domestic politics, law, and practices are in this field. The examples of these contradictions are legion:

- At the same time that Google was engaged in high-stakes negotiations with China over how to censor search results returned in China, thousands of specially equipped vehicles were driving through the streets of the United States and Europe collecting detailed imagery and telemetry to help improve Google's online services and business offerings; the legality of this practice is still being debated in the United States and abroad.[26]

- In February 2011, at the same time that the secretary of state was delivering a major speech on Internet freedom,[27] the Department of Justice was pursuing a grand jury criminal investigation, at the express direction of the attorney general, of Julian Assange over the release on the Internet of classified U.S. diplomatic cables.[28]

The point is that it is difficult if not impossible to have a coherent outward-facing policy when there is no coherent, settled inward-facing policy.

Second, it is infeasible for the United States to have a general policy on Internet freedom because the instruments of repression on the Internet are indistinguishable from the instruments used by governments worldwide to accomplish socially acceptable aims. The relevant differences have to do with the purposes and procedures of government intervention in the Internet, the validity of which are entirely subjective.

What the United States can and should do is insert specific aspects of Internet freedom into bilateral foreign relationships after taking U.S. law and practices in this area fully into account and while considering the broader, strategic context of the relationship. For instance, it probably makes sense for the U.S. government to provide technical assistance aimed at helping dissident groups circumvent the specific Internet censorship practices of a repressive government with whom the U.S. government has overtly hostile relations. In bilateral relations short of outright hostility, however, anti-circumvention assistance to dissidents is essentially a policy that aims to subvert a sovereign government with whom the United States is otherwise constructively engaged.

Third, with respect to U.S.-based technology companies being pressured by governments into becoming agents of Internet repression, if there is any general policy that makes sense, it is probably to keep out. This is a true conundrum, because it occasionally puts these private entities in the middle of high-stakes geopolitics, like the "Arab Spring" or coping with the rise of Chinese power, where a technology corporation's decision can have real bearing on a range of U.S. national interests. But on the other side of the ledger is the fact that sovereign governments have a right, within their own borders, to regulate commerce as they see fit. The United States certainly has not been shy about exercising this prerogative. There undoubtedly will be situations in which it makes sense for the U.S. government to intervene—consider, for instance, the State Department's request in June 2009 that Twitter not take its service offline for routine maintenance at a critical moment in the Iranian citizens' protests over the reelection of President Mahmoud Ahmadinejad. But these cases should be exceptional, carefully considered, and fact-dependent.

Richard Falkenrath is Principal at the Chertoff Group, Contributing Editor at Bloomberg News, and Adjunct Senior Fellow at the Council on Foreign Relations. Previously, he was the Deputy Commissioner of Counterterrorism at the New York City Policy Department; Deputy Assistant to the President and Deputy Homeland Security Advisor; member of the National Security Council Staff; and Assistant Professor of Public Policy at the John F. Kennedy School of Government. He received a Ph.D. from the Department of War Studies, Kings College London, where he was a British Marshall Scholar; and a *summa cum laude* B.A. from Occidental College in economics and international affairs. He is a member of the Aspen Strategy Group.

[1] Secretary of State Hillary Clinton, "Internet Freedom" (remarks delivered at The Newseum, Washington, D.C., January 21, 2010), http://www.state.gov/secretary/rm/2010/01/135519.htm.

[2] U.S. President Barack Obama (remarks delivered at the Town Hall Meeting with Future Chinese Leaders, Museum of Science and Technology, Shanghai, China, November 16, 2009), http://www.whitehouse.gov/the-press-office/remarks-president-barack-obama-town-hall-meeting-with-future-chinese-leaders.

[3] Barack Obama, "On Securing Our Nation's Cyber Infrastructure" (remarks delivered in the East Room, The White House, May 29, 2009), http://www.whitehouse.gov/the_press_office/Remarks-by-the-President-on-Securing-Our-Nations-Cyber-Infrastructure/.

[4] "Internet freedom" in this paper refers to information moved on a network—wired or wireless—by any electronic media, including computers, smart phones, mobile devices, and cloud services.

[5] See Evgeny Morozov, "Why the Internet Is a Great Tool for Totalitarians," *Wired*, January 2010, http://www.wired.com/magazine/2010/12/st_essay_totalitarians/.

[6] Kim Zetter, "Judge Won't Stop WikiLeaks Twitter-Records Request," *Wired*, April 2011, http://www.wired.com/threatlevel/2011/03/judge-denies-on-twitter-case/.

[7] Deputy Assistant Attorney General, Criminal Division Jason Weinstein "Data Retention as a Tool For Investigating Internet Child Pornography and Other Internet Crimes" (statement before the Committee

on Judiciary, Subcommittee on Crime, Terrorism, and Homeland Security, United States House of Representatives, Washington, D.C., January 25, 2011).

[8] Charlie Savage, "As Online Communications Stymie Wiretaps, Lawmakers Debate Solutions," *New York Times*, February 17, 2011, http://www.nytimes.com/2011/02/18/us/18wiretap.html.

[9] Christopher Rhoads, Geoffrey A. Fowler and Chip Cummins, "Iran Cracks Down on Internet Use, Foreign Media," *Wall Street Journal*, June 17, 2009, http://online.wsj.com/article/SB124519888117821213.html.

[10] Greg Sandoval, "Exclusive: Top ISPs poised to adopt graduated response to piracy," *CNET News*, June 22, 2011, http://news.cnet.com/8301-31001_3-20073522-261/exclusive-top-isps-poised-to-adopt-graduated-response-to-piracy/.

[11] For an excellent overview of censorship technology, see Ronald Deibert, et al., eds., *Access Denied: The Practice and Policy of Global internet Filtering* (Cambridge: MIT Press, 2008).

[12] Filtering packet-based Internet transmissions by content—for example, by using a proxy server to reassemble the packets of binary data that make up an Internet transmission, or by packet inspection—is a more difficult and computationally demanding task, as discussed below.

[13] Except, generally, to Iran, Sudan, and Cuba, pursuant to the United States Office of Foreign Assets Control (OFAC) sanction list.

[14] The term "Censorship 2.0" comes from Robert Faris, Stephanie Wang, and John G. Palfrey, "Censorship 2.0," *Innovations* (Spring 2008).

[15] "International: Fingered; The War on WikiLeaks." *The Economist*, December 11, 2010, 71-72, http://www.economist.com/node/17674107.

[16] Letter from U.S. Department of State Legal Advisor Harold Koh to Ms. Jennifer Robinson, attorney for Mr. Julian Assange, November 27, 2010.

[17] For excellent discussion of these issues, see Jonathan Zittrain and John Palfrey, "Reluctant Gatekeepers: Corporate Ethics on a Filtered Internet," in Deibert, et al., *Access Denied*.

[18] Recall the operatic opening lines of *A Declaration of the Independence of Cyberspace* delivered at Davos in 1996: "Governments of the Industrial World, you weary giants of flesh and steel, I come from Cyberspace, the new home of Mind. On behalf of the future, I ask you of the past to leave us alone. You are not welcome among us. You have no sovereignty where we gather."

[19] This is the central theme of Jack Goldsmith and Tim Wu, *Who Controls the Internet? Illusions of a Borderless World* (New York: Oxford University Press, 2006).

[20] Glanz, James and John Markoff, "Vast Hacking by a China Fearful of the Web," *The New York Times*, March 10, 2010, http://www.nytimes.com/2010/12/05/world/asia/05wikileakschina.html?adxnnl=1&adxnnlx=1310748334-BH5ATwnRDK8MpkE8HK3xHQ.

[21] Miriam Marcus, "Google and 10 Chinese Web Brothers," *Forbes.com*, March 29, 2010, http://www.forbes.com/2010/03/29/google-china-baidu-intelligent-investing-ctrip-microsoft.html.

[22] A more advanced, computationally demanding strategy involves using computers between the sender and the recipient (called "proxy servers") to reassemble the packets into coherent files, filter the content in these files, and retransmit only those files that meet pre-determined criteria.

[23] Ellen Nakashima, "Pentagon Considers Preemptive Strikes as Part of Cyber-Defense Strategy," *The Washington Post*, August 23, 2010, http://www.washingtonpost.com/wp-dyn/content/article/2010/08/28/AR2010082803849.html.

[24] Sean Hansell, "The Economics of Snooping on Internet Traffic." *The New York Times*. March 25, 2009, http://bits.blogs.nytimes.com/2009/03/25/the-economics-of-snooping-on-internet-traffic/?scp=2&sq=deep%20packet%20inspection&st=cse.

[25] See sidebar, "Mixed Signals" in the April 2011 Middle East Notes and Comment bulletin from the Center for Strategic and International Studies, http://csis.org/files/publication/0411_MENC2.pdf.

[26] "The World on Your Desktop," *The Economist*, September 8, 2007, http://www.economist.com/node/9719045.

[27] Secretary of State Hillary Clinton, "Internet Rights and Wrongs: Choices and Challenges in a Networked World" (remarks delivered at George Washington University, Washington, D.C., February 15, 2011), http://www.state.gov/secretary/rm/2011/02/156619.htm.

[28] Julian Barnes and Evan Perez, "Assange Probe Hits Snag," *The Wall Street Journal*, February 9, 2011, http://online.wsj.com/article/SB10001424052748703313304576132543747598766.html.

"...the local political context is critical. The medium may be global, but whether and how it enables individuals to foster democratic change depends to a large degree on a wide array of local variables, including opposition leadership, the existence of civil society institutions, the willingness of the regime to crack down on dissident activity, and so forth."

—RICHARD FONTAINE

Internet Freedom and Political Change*

Richard Fontaine
Senior Advisor
Center for a New American Security

The Internet's potential as a tool for political change exploded onto the radar screens of top foreign policy officials in 2009, with what was quickly dubbed Iran's "Twitter Revolution." Protestors used the Internet and text messages to spread information and coordinate efforts, and the potential of new technologies was crystallized in the viral movement of a video depicting the brutal slaying of a young Iranian student, Neda Agha-Soltan. The video, which was captured on a mobile phone and uploaded to YouTube, traveled across the Web and onto local and satellite television, prompting President Obama to express his outrage at the killing. When the president of the United States uses a White House press conference to address material uploaded to YouTube, something fundamental has changed in the nature of modern communications.

The focus on the role played by new communications technologies grew as the Arab Spring gathered momentum a year and a half later. The wave of revolts across the Arab world that swept across Tunisia, Egypt, Libya, Bahrain, Yemen, Syria, and elsewhere was fueled in part by activists using tools such as Facebook, Twitter, SMS (text messaging), and other digital platforms. Several regimes took draconian steps to stop online organizing and communication; Egypt cut off Internet access to the entire country for five days. Among observers, the sense grew that the Internet matters to these dramatic political shifts, but just how it matters was not entirely clear.

In a sense, the Internet represents just the latest part of a story that has unfolded for centuries. Communication technologies have played significant roles in political movements since antiquity; from the printing press that empowered the Protestant Reformation, to the cassette tapes distributed by revolutionaries throughout Iran in 1979, to the use of fax machines by Poland's Solidarity movement, and the effect of satellite television today. But the global nature of the Internet, its very low barrier to

---

* *This paper is adapted in part from portions of Richard Fontaine and Will Rogers, "Internet Freedom: A Foreign Policy Imperative in the Digital Age," Center for a New American Security, June 2011.*

entry, its speed, and the degree to which it empowers the individual have introduced a new force into political discourse and mobilization. The Internet itself has become the focus of attention of dictators, democracy activists, and observers around the world.

Does Internet Freedom Lead to Democracy?

The United States promotes Internet freedom for two reasons. First, Americans tend to believe in the freedom of expression, in any medium. Second, American policymakers have bet that on balance, the increased availability of new, unfettered communications technologies abets the spread of democracy. But does it?

Experts remain deeply divided, as the quotations below illustrate, as to whether unbridled access to the Internet can help transform authoritarian regimes over time and bring greater freedom to once-closed societies. Most of those who have attempted to assess its impact have relied on case studies, anecdotes, or theory. Because of the novelty of the phenomenon and the few and widely varying data points, there are notable analytical challenges. In addition, much of the assessment is a matter of subjective interpretation. Facebook clearly played a major role in building an opposition to the Mubarak regime and in organizing protests.[1] But after Egypt shut off the Internet, protests became bigger, not smaller. Did this demonstrate the Internet's limited role as a tool of agitation? Or did the shutoff of cherished online tools itself spur enraged citizens to demonstrate instead of staying home (possibly in front of a computer)?

Does the Internet Promote Democracy?

Optimists

"The Internet is above all the most fantastic means of breaking down the walls that close us off from one another. For the oppressed peoples of the world, the Internet provides power beyond their wildest hopes."[2]

⇨ **Bernard Kouchner**, former French foreign minister

"It does make a difference when people inside closed regimes get access to information—which is why dictatorships make such efforts to block comprehensive Internet access... [promoting Internet freedom] would be a cheap and effective way of standing with Iranians while chipping away at the 21st-century walls of dictatorship."[3]

⇨ **Nicholas Kristof**, *New York Times* columnist

"The Internet is possibly one of the greatest tools for democratization and individual freedom that we've ever seen."[4]

⇨ **Condoleezza Rice**, former secretary of state

"Without Twitter, the people of Iran would not have felt empowered and confident to stand up for freedom and democracy."[5]

⇨ **Mark Pfeifle**, former deputy national security advisor

"If you want to liberate a society, just give them the Internet."[6]

⇨ **Wael Ghonim**, Egyptian Google executive and democracy activist

Skeptics

"The idea that the Internet favors the oppressed rather than the oppressor is marred by what I call cyber-utopianism: a naïve belief in the emancipatory nature of online communication that rests on a stubborn refusal to admit its downside."[7]

⇨ **Evgeny Morozov**, author of *The Net Delusion*

"The platforms of social media are built around weak ties . . . weak ties seldom lead to high-risk activism."[8]

⇨ **Malcolm Gladwell**, *New Yorker* staff writer

"Democracy isn't just a tweet away."[9]

⇨ **Jeffrey Gedmin**, former president of Radio Free Europe / Radio Liberty

"It is time to get Twitter's role in the events in Iran right. Simply put: There was no Twitter Revolution inside Iran."[10]

⇨ **Golnaz Esfandiari**, senior correspondent for Radio Free Europe / Radio Liberty

"Techno-optimists appear to ignore the fact that these tools are value neutral; there is nothing inherently pro-democratic about them. To use them is to exercise a form of freedom, but it is not necessarily a freedom that promotes the freedom of others."[11]

⇨ **Ian Bremmer**, president of the Eurasia Group

It has become axiomatic to say that the Internet does not itself create democracies or overthrow regimes; people do. This is obviously true, but if new communications tools do matter—and there appears to be at least nascent evidence that they do—then they can play a role in several distinct ways. An important report issued by the United States Institute of Peace (USIP) presents a useful framework for examining how new communications technologies might affect political action. The report identifies five distinct mechanisms through which the Internet might promote (or be used by regimes to retard) democratic progress.[12] This chapter deepens the analysis of these mechanisms and adds two additional factors that affect them.

The Internet may affect individuals, by altering or reinforcing their political attitudes, making them more attuned to political events, and enabling them to participate in politics to a greater degree than they could otherwise. This does not automatically translate into a more activist population; as the USIP study notes, it could actually make citizens more passive by diverting their attention away from offline political activism and toward insignificant online activity.[13] Some have called this "slacktivism," the quintessential example of which is millions of individuals signing online petitions to end genocide in Darfur but taking no further action.[14] At the same time, individuals freely expressing themselves on the Internet are exercising a basic democratic right. As democracy scholar Larry Diamond points out, used in this way, the Internet can help "widen the public sphere, creating a more pluralistic and autonomous arena of news, commentary, and information."[15] It can also serve as an instrument through which individuals can push for transparency and government accountability, both of which are hallmarks of mature democracies.[16]

New media might also affect intergroup relations, by generating new connections among individuals, spreading information, and bringing together people and groups. (Some have worried about the opposite effect—the tendency of the Internet to polarize individuals and groups around particular ideological tendencies.)[17] This may occur not only within countries but among them; the protests in Tunisia sparked a clear rise in political consciousness and activism across the Arab world—much of it facilitated by Internet-based communications and satellite television.[18] It may also take place over a long period of time; Clay Shirky, an expert at New York University, argues that a "densifying of the public sphere" may have to take place before an uprising turns into a revolution.[19]

New communications technologies could also affect collective action, by making it easier for individuals and groups to organize protests and change opinions in a repressive country. Unconnected individuals dissatisfied with the prevailing politics may realize that others share their views, and this might form the basis for collective action.[20] Relatively small groups, elites, or other motivated dissidents might use the Internet to communicate or organize protests. Even if the number of committed online activists is small, they might nevertheless disseminate information to the general population or inspire more widespread protests.[21] Again, it is important to distinguish such action from group "slacktivism"—as the successful protests in Egypt showed, only once thousands of citizens physically occupied Tahrir Square did the regime begin to teeter. Though at the outset the protests may have been organized via Facebook, had they been confined only to cyberspace, the Mubarak government would still be in power.

These new technologies clearly affect regime policies as well. Governments have employed a huge array of techniques aimed at controlling the Internet and ensuring that their political opponents cannot use it freely. This goes well beyond censorship, which garners the bulk of popular attention. Autocracies also regularly monitor dissident communications, mobilize regime defenders, spread propaganda and false information designed to disrupt protests and outside groups, infiltrate social movements, and disable dissident websites, communications tools, and databases. These and other practices can also induce self-censorship and other forms of self-restraint by publishers, activists, online commentators, and opposition politicians.

Autocrats can also turn dissidents' use of the Internet against them. In Iran, for example, users of social media—which linked their accounts to those of other protestors—inadvertently created a virtual catalogue of political opponents through which the government could identify and prosecute individuals. The regime established a website that published the photos of protestors and used crowd-sourcing to identify the individuals' names.[22] Similarly, the Revolutionary Guard reportedly sent intimidating messages to those who posted pro-opposition messages and forced some citizens entering the country to open their Facebook accounts upon arrival.[23] In the midst of the Arab Spring protests, Syria allowed its citizens to access Facebook and YouTube for the first time in three years; some human rights activists suspected that the government made the change precisely in order to monitor people and activities on these sites.[24]

Similarly, shortly after the Egyptian government lifted its Internet blackout in early 2011, pro-Mubarak supporters conducted an online disinformation campaign, using Facebook and Twitter to disrupt planned demonstrations by posting messages saying the protests had been canceled.[25] The government reportedly sent Facebook messages to citizens urging them not to attend protests because doing so would harm the Egyptian economy.[26] In the same vein, the Chinese government employs an estimated 250,000 "50 Cent Party" members who are paid a small sum each time they post a pro-government message online.[27] And after an anonymous post on the U.S.-based Chinese language website Boxun.com called on activists to stage China's own "Jasmine Revolution," no demonstrators turned up at the rally point—instead it was flooded with security teams and plainclothes officers.[28] Some speculated that Chinese officials themselves may have authored the anonymous posting in an effort to draw out political dissidents.[29] While no evidence has emerged to support the claim, it is not hard to imagine such an attempt taking place in the future.

Autocracies are engaged in "offline" attempts to repress Internet use as well. Saudi Arabia, for example, has not only blocked websites but also placed hidden cameras aimed at monitoring user behavior in Internet cafés and required café owners to make their customer lists available to government officials.[30] China requires users to register their identification upon entry to a cybercafé.[31] And Libyan officials simply demanded that refugees fleeing the recent fighting turn over their cell phones or SIM cards at border checkpoints.[32]

Beyond these effects, new media can draw external attention, by transmitting images and information to the outside world beyond the control of government-run media and regime censorship and spin. Such attention can mobilize sympathy for protestors or hostility toward repressive regimes.[33] This was exemplified by the way the video of Neda Agha-Soltan moved from YouTube to mainstream media. Digital videos and information may also have a rebound effect within the country in question; information transmitted out of Egypt and Libya by social networking and video-hosting sites during the protests in those countries made their way back in via widely-watched satellite broadcasts. This effect could be particularly pronounced in countries like Yemen, where Internet penetration is low but Al Jazeera is widely viewed. Similarly, print journalists have found sources and stories through social media and have used the same media to push their articles out to the world.

The economic impact of the Internet might also affect the degree of democratization in a country. The Internet has generated an increase in labor productivity and corresponding economic growth,[34] which may help middle classes emerge in developing countries. Because new middle classes tend to agitate for democratic rights, new technologies could produce an indirect democratizing effect. Secretary of State Clinton referenced a related dynamic, the "dictator's dilemma," in a 2011 speech, stating that autocrats "will have to choose between letting the walls fall or paying the price to keep them standing . . . by resorting to greater oppression and enduring the escalating opportunity cost of missing out on the ideas that have been blocked and people who have been disappeared."[35] In other words, an autocrat can either repress the Internet or enjoy its full economic benefits, but not both.

Whether the "dictator's dilemma" actually exists is yet unknown. There are certainly clear individual instances where Internet repression has damaged a nation's economy: OECD experts have estimated that Egypt's five-day Internet shutdown cost the country at least $90 million, a figure that did not include e-commerce, tourism or other businesses that rely on Internet connectivity.[36] But China seems to provide a powerful counterexample, since it severely represses the Internet while enjoying extraordinarily high rates of sustained economic growth. Indeed, China appears to have used its restrictive Internet practices to squeeze out international competition and generate the conditions under which only domestic companies—ones that adhere to the stringent censorship and monitoring practices employed by the state—can thrive. China's largest domestic search engine, Baidu, exercises strict controls on content, but has thrived since Google pulled out of the country in January 2010. China may be an outlier; the massive financial and human resources it devotes to online control may not be replicable elsewhere. Other countries may be left with blunter forms of repression that degrade both the Internet's economic and political effects.

New technologies can also accelerate the political and economic effects described above. Google's Eric Schmidt and Jared Cohen have argued that faster computer power combined with the "many to many" geometry of social media empowers individuals and groups at the expense of governments and that this, in turn, increases the rate of change.[37] Dissidents can identify one another, share information, organize, and connect with leaders and with external actors, all more easily and faster than ever before.[38] Indeed, one hallmark of the 2011 Arab Spring was the astonishing rate of change and the ability of popular protests to threaten or topple governments that had been in power for decades in a matter of weeks.[39]

Again, the local political context is critical. The medium may be global, but whether and how it enables individuals to foster democratic change depends to a large degree on a wide array of local variables, including opposition leadership, the existence of civil society institutions, the willingness of the regime to crack down on dissident activity, and so forth. Take, for example, the call for political change during the Arab Spring. In Tunisia and Egypt, tens of thousands of protestors responded to protest event pages on Facebook by taking to the streets. Yet in other Arab states a call on Facebook for a "day of rage" did not have the same pronounced influence. The degree of openness in the local political system, the discontent among the population, the willingness of the government to use coercive means to stop democratic activism, the role of minorities, and much more play a key role.

While there is no absolute indication that the Internet will engender the democratization of societies that many hope for, some tentative conclusions are warranted. First, it is clear that there is no determinism: Iran saw a "Twitter Revolution" that spurred no emancipation; Egypt saw a "Facebook Revolution" that toppled the Mubarak regime. Second, the technology itself is agnostic, and the same online tools that empower dissidents can aid dictators in their oppression. In the short run, at least, a freer Internet does not automatically translate into more liberal political systems.

Some of the case studies do, however, demonstrate the Internet's profound potential: access to an open Internet can help countries slide away from authoritarianism and toward democracy. Events in places like Iran, Tunisia, Egypt, and elsewhere suggest that the Internet and related technologies (such as SMS) have indeed served as critical tools for organizing protests, spreading information among dissident parties, and transmitting images and information to the outside world— some of which moved onto satellite television channels, which further boosted their influence.[40] And while experts continue to argue about the precise effect, they tend to agree that social media tools have made revolution in the Middle East easier and speedier than it would have otherwise been.[41]

Perhaps the most compelling reason to see the democratizing potential of a free Internet is also the simplest: Both dissidents and dictatorships abroad seem to believe that the Internet can have a transformative role, and they act on that basis. Dictatorships expend enormous time and resources to clamp down on online activity, and more than forty countries actively censor the Internet or engage in other forms of significant Internet repression.[42] Meanwhile, millions of individuals use proxy servers and other circumvention and anonymity tools to evade censorship and monitoring. To cite one example, during the 2009 presidential campaign in Iran, both President Mahmoud Ahmadinejad and his opponent Mir-Hussein Mousavi cited the Internet as a tool through which the liberal opposition could mobilize support.[43] It is unlikely they were both wrong. While the effect of the Internet will depend on local conditions, there are indeed reasonable grounds for believing that a free Internet can play an important role in empowering individuals to move toward more liberal political systems.

The U.S. Government Role

The U.S. government promotes Internet freedom in five main ways: providing Internet technologies, shaping international norms to favor the free flow of online information, encouraging the private sector to expand its role in bolstering that flow of information, using economic diplomacy to persuade other states to embrace an open Internet, and reforming export controls to permit users in Internet-repressing states to access U.S.-made software and technologies that could allow them to communicate freely and securely. The following section will touch briefly on the government's provision of Internet technologies and then on the potential to use trade policy in support of online freedom.

Supporting technologies

In the face of autocracies' attempts to censor, identify, intimidate, and monitor online users, the U.S. government has focused on providing technology that allows individuals living in repressive environments to freely access online information. The private sector has few financial incentives to develop such technologies—it is difficult to charge anonymous subscribers or sell ads for such services in closed societies—and very few foreign governments, nongovernmental organizations, or foundations have yet funded them on their own.

The State Department has spent approximately $20 million since 2008 on programs to develop circumvention technologies and promote digital activism, and plans to award more than $25 million in additional funding in 2011.[44] In April 2011, Congress reallocated $10 million from the State Department to the Broadcasting Board of Governors (BBG)—an amount that reduced State's Internet freedom budget by a third and more than quintupled the BBG's budget in this area. The Defense Advanced Research Projects Agency (DARPA) currently funds the development of circumvention technologies that would allow the U.S. military to safely and anonymously access the Internet.[45]

A variety of circumvention technologies enable dissidents to penetrate firewalls and access blocked websites and censored information. Each tool employs the same basic method: it routes a user's request through an unblocked webpage in order to access banned content. A user in China, for instance, who cannot access the *New York Times* website could, instead, reach a proxy site that could then obtain the *Times'* web content.

Other technologies help users maintain their anonymity in the face of a regime's watchful eye. One notable tool has been made available by the Tor Project, which received nearly $750,000 from the U.S. government between 2006 and 2010.[46] Tor uses a network in which encrypted messages pass through several nodes known as "onion routers" that then peel away layers of encryption as information is transmitted among proxy servers around the world. The network allows users to hide their location from websites they are visiting, enabling them to evade governments and others attempting to trace their location. In addition, virtual private networks (VPNs) encrypt and tunnel all Internet traffic through a proxy, enabling their users to circumvent firewalls and use webmail, chat and other online communication services.

Other technological tools enhance the ability of dissidents and activists to freely use the online space. Software exists to help protect websites against denial of service attacks, which can be launched by autocratic regimes or patriotic hackers (individuals or groups who express nationalistic pride by attacking foreign government or dissident websites) by sending millions of page requests per second to a site, thereby overloading and crashing its servers. Other available tools can help secure online databases (of human rights abuses, for example), provide mirror sites to keep websites live during an attack, and archive uploaded data so that it can be easily reposted after a website returns to service. In addition, as mobile technology increasingly becomes a main platform for online activity, there is increasing interest in secure cell phones and encrypted communications.

There are dilemmas associated with providing such technologies. They cannot be used effectively by activists who lack the skills to employ them, and they can actually be dangerous. Used improperly, they may give users a false sense of security or expose their identities and online actions to authorities. In addition, funding anonymity technologies may conflict with cybersecurity impulses, which emphasize the need for online attribution. The Tor network, for example, does not have a back door through which the U.S. government or other law enforcement agencies can access and monitor the secured communication or web traffic. As a result, it is possible that these technologies could be used not only by dissidents and democracy activists, but also by criminals and terrorists.[47] While not reacting specifically to government-funded anonymity tools, the FBI has been outspoken for years about the potential risks associated with the spread of sophisticated encryption technologies.[48]

It is important to note, however, that U.S. government-supplied circumvention tools are not the only option for individuals wishing to communicate anonymously or access banned websites. Criminals and terrorists are far more likely to use botnets (collections of compromised computers running automated software, generally without the knowledge of their users) and other illicit tools instead of settling for the less effective tools offered by the U.S. government. (The government-sponsored tools can be slower than others, have restricted bandwidth, and contain other features that make illicit tools more attractive by comparison.) "Mujahideen Secrets 2," for example, is a jihadi-developed encryption tool designed to allow al Qaeda supporters to communicate online.[49] While it is clearly impossible to eliminate the possibility that government-sponsored technologies will be used by bad actors, it is likely that those numbers will pale in comparison to the quantity of users simply wishing to access neutral media.

Trade policy

Though it has received relatively little attention thus far, trade policy provides one potential avenue for the U.S. government to promote Internet freedom. Internet repression serves as a trade barrier; when a country blocks access to a U.S. website, for example, it also blocks the site's advertising—and thereby interferes with the trade in products and services advertised.[50] Of the millions of dollars lost during the Internet shutoff in Egypt, it is hard to imagine that none would have accrued to American businesses.

Employing trade agreements has an advantage over other forms of diplomatic persuasion in that they contain economic incentives (and thus give the United States negotiating leverage) and are at least potentially enforceable. Should Internet censorship become accepted as a non-tariff trade barrier, a censoring government could be vulnerable to dispute arbitration at the World Trade Organization or to bilateral trade remedies. And such agreements could be bilateral, multilateral or even global.

The Korea-U.S. Free Trade Agreement contains a relevant provision: "Recognizing the importance of the free flow of information in facilitating trade, and acknowledging the importance of protecting personal information, the Parties shall endeavor to refrain from imposing or maintaining unnecessary barriers to electronic information flows across borders."[51] Such language is clearly nonbinding—"shall endeavor to refrain" is a loose commitment at best—but nevertheless suggests how the United States can promote Internet freedom in future trade negotiations.

Conclusion

Over the past several years, the U.S. government has taken important, positive steps to promote Internet freedom in a number of areas, ranging from providing technologies to shaping norms to engaging with the private sector. Given the effects this chapter describes above, the government should now build on these efforts to integrate other elements, including using trade policy to promote the free flow of information, modifying export controls to permit the provision of certain technologies to repressive states, and others. Underlying all these efforts is a bet—essentially the same bet that the United States placed during the Cold War—that supporting access to information and encouraging the free exchange of ideas is good for America. That bet is well worth making.

A free Internet, however, does not represent some silver bullet for social change. Supporting Internet freedom is complicated and poses tradeoffs with other items on the American diplomatic, security, and economic agenda. It should be seen as just one, potentially quite important, element in a broader approach to promoting ideas and ideals in repressive societies. The net effect of this effort is uncertain, and it will likely remain so for years.

But we should not underestimate the potential power of the Internet. We live in a time when an application like Facebook—designed in 2004 for American university students to share information—has, in 2011, been used to help topple a dictator in Egypt; a time when the best satellite television coverage of demonstrations and conflict can come from online video postings; and a time when dissidents risk imprisonment or worse for blogging their beliefs.

The U.S. government faces a constant challenge in keeping up with new technology and the changing ways that users employ it. Corporations are continually vexed by the many varying demands put upon them by governments around the world. Individuals in autocratic societies face dilemmas in determining how to proceed online. Though the debate is complicated, the longstanding American commitment to basic human rights and freedoms should remain clear. And on that basis, the United States has a responsibility to promote Internet freedom, the key to ensuring a greater degree of human liberty in an ever more contested space.

Richard Fontaine is a Senior Advisor at the Center for a New American Security. He is a member of the Council on Foreign Relations and is an adjunct professor in the Security Studies Program at Georgetown University's School of Foreign Service. He previously served as foreign policy advisor to Senator John McCain for more than five years. He has also worked at the State Department, the National Security Council and on the staff of the Senate Foreign Relations Committee. Mr. Fontaine graduated summa cum laude with a B.A. in International Relations from Tulane University. He also holds an M.A. in International Affairs from the Johns Hopkins School of Advanced International Studies (SAIS) in Washington, and he attended Oxford University.

[1] At the beginning of 2011, more than a fifth of the Egyptian population used the Internet and approximately five million Egyptians had Facebook accounts. See: Internet World Stats: Usage and Population Statistics: http://www.internetworldstats.com/af/eg.htm and Jennifer Preston, "Facebook and YouTube Fuel the Egyptian Protests," *New York Times*, February 5, 2011. Available at http://www.nytimes.com/2011/02/06/world/middleeast/06face.html?_r=1&pagewanted=all.

[2] Bernard Kouchner, "The Battle for the Internet," *New York Times*, May 13, 2010. Available at http://www.nytimes.com/2010/05/14/opinion/14iht-edkouchner.html?_r=1.

[3] Nicholas D. Kristof, "Tear Down This Cyberwall!" *New York Times*, June 17, 2009. Available at http://www.nytimes.com/2009/06/18/opinion/18kristof.html?_r=1.

[4] Under Secretary of State for Democracy and Global Affairs Paula Dobriansky, "New Media vs. New Censorship: The Assualt" (Remarks delivered to the Broadcasting Board of Governors, Washington, D.C., September 10, 2008).

[5] Mark Pfeifle, "A Nobel Peace Prize for Twitter?" *Christian Science Monitor*, July 6, 2009. Available at http://www.csmonitor.com/Commentary/Opinion/2009/0706/p09s02-coop.html.

[6] Alexel Oreskovic, "Egyptian Activist Creates Image Issue for Google," *Reuters*, February 12, 2011. Available at http://www.reuters.com/article/2011/02/12/us-egypt-google-idUSTRE71B0KQ20110212.

[7] Evgeny Morozov, *The Net Delusion: The Dark Side of Internet Freedom* (Washington, DC: PublicAffairs, 2011), xiii.

[8] Malcom Gladwell, "Why the Revolution Won't be Tweeted," *The New Yorker*, October 4, 2010. Available at http://www.newyorker.com/reporting/2010/10/04/101004fa_fact_gladwell?currentPage=1.

[9] Jeffrey Gedmin, "Democracy isn't Just a Tweet Away," *USA Today*, April 22, 2010.

[10] Golnaz Esfandiari, "Misreading Tehran: The Twitter Devolution," *Foreign Policy*, June 7, 2010. Available at http://www.foreignpolicy.com/articles/2010/06/07/the_twitter_revolution_that_wasnt.

[11] Ian Bremmer, "Democracy in Cyberspace," *Foreign Affairs* (November/December 2010). Available at http://www.foreignaffairs.com/articles/66803/ian-bremmer/democracy-in-cyberspace.

[12] Sean Aday, Henry Farrell, Marc Lynch, John Sides, John Kelly, and Ethan Zuckerman, "Blogs and Bullets: New Media in Contentious Politics," United States Institute of Peace (August 2010): 9. Much of the discussion of theories of change in this section draws upon this important work.

[13] Ibid., 9.

[14] Evgeny Morozov, "From Slacktivism to Activism," *Foreign Policy Net Effect* blog, September 5, 2009. Available at http://neteffect.foreignpolicy.com/posts/2009/09/05/from_slacktivism_to_activism.

[15] Larry Diamond, "Liberation Technology," *Journal of Democracy* (July 2010): 70.

[16] Ibid.

[17] Aday et. al., "Blogs and Bullets": 10.

[18] Steve Coll, "The Internet: For Better or for Worse," *The New York Review of Books*, April 7, 2011.

[19] Lauren Kirchner, "'Information Wars' on Al Jazeera English," *Columbia Journalism Review*, February 14, 2011.

[20] Ibid., 10-11. The authors point out that social media in particular may reduce the transaction costs for organizing collective action, for example by making communication easier across physical and social distance, or by undermining top-down movements in favor of flatter social movements.

[21] Ibid., 18.

[22] It is worth noting that this website has evidently garnered little interest. Evgeny Morozov, "Think Again: The Internet," *Foreign Policy* (May/June 2010). Available at http://www.foreignpolicy.com/articles/2010/04/26/think_again_the_internet.

[23] "'Haystack' Gives Iranian Opposition Hope for Evading Internet Censorship," *Christian Science Monitor*, April 16, 2010.

[24] Jennifer Preston, "Syria Restores Access to Facebook and YouTube," *New York Times*, February 9, 2011.

[25] Spencer Ackerman, "Trolls Pounce on Facebook's Tahrir Square," *Wired Magazine* (February 2011). Available at http://www.wired.com/dangerroom/2011/02/trolls-pounce-on-facebooks-tahrir-square/.

[26] Tarek Amr, "The Middle East, the Revolution, and the Internet" (Remarks at AccessNow web symposium, February 3, 2011).

[27] Daniel Calingaert, "Authoritarianism vs. the Internet," *Policy Review* (April 2010): 6.

[28] Brianna Lee, "Chinese Government Issues Preemptive Crackdown of 'Jasmine Revolution' Protests," *PBS.org*, March 3, 2011. Available at http://www.pbs.org/wnet/need-to-know/the-daily-need/chinese-government-issues-preemptive-crackdown-of-jasmine-revolution-protests/7697/.

[29] Quincy Yu, "Aborted Chinese 'Jasmine Revolution' a Trap Say Analysts," *The Epoch Times*, February 22, 2011. Available at http://www.theepochtimes.com/n2/china/aborted-chinese-jasmine-revolution-a-trap-say-analysts-51732.html.

[30] 2010 Reporters Without Borders report cited in "New Media: A Force for Good or Evil?" *The Layalina Review* (12-25 March 2010).

[31] Rebecca MacKinnon, "China, the Internet, and Google," (Testimony before the Congressional-Executive Commission on China, Washington, D.C., March 1, 2010), 7.

[32] Scott Peterson, "On Libya-Tunisia Border, Refugees Plead for Help to Go Home," *Christian Science Monitor*, March 3, 2011.

[33] Ibid., 12.

[34] The Organisation for Economic Co-operation and Development (OECD) published a 2007 study on the economic impact of broadband Internet access based on the work done by the Working Party on the Information Economy. The study notes that information and communications technologies (ICTs) enable

measurable economic growth: "Broadband is also increasingly important as an enabling technology for structural changes in the economy, most notably via its impact on productivity growth, but also by raising product market competition in many sectors, especially in services. ICTs and broadband are facilitating the globalisation of many services, with broadband making it feasible for producers and consumers of services to be in different geographical locations." OECD Ministerial Background Report, *Broadband and the Economy* (May 2007). Available at http://www.oecd.org/dataoecd/62/7/40781696.pdf.

[35] Secretary of State Hillary Rodham Clinton, "Internets Rights and Wrongs: Choices & Challenges in a Networked World" (remarks given at George Washington University, Washington, D.C., February 15, 2011). Available at http://www.state.gov/secretary/rm/2011/02/156619.htm.

[36] "The Economic Impact of Shutting Down Internet and Mobile Phone Services in Egypt," OECD Report, February 4, 2011. Available at http://www.oecd.org/document/19/0,3746,en_2649_34223_47056659_1_1_1_1,00.html.

[37] Eric Schmidt and Jared Cohen, "The Digital Disruption: Connectivity and the Diffusion of Power," *Foreign Affairs* (November/December 2010).

[38] Doyle McManus, "Did Tweeting Topple Tunisia?" *Los Angeles Times*, January 23, 2011.

[39] As is true when evaluating the effect of new technologies along other dimensions of political change, here too it is important to leaven exuberance with caution. Historian Simon Sebag Montefiore, for instance, notes that while Facebook "certainly accelerates the mobilization of crowds. . . technology's effect is exaggerated: in 1848, the revolution that most resembles today's, uprisings spread from Sicily to Paris, Berlin, Vienna and Budapest in mere weeks without telephones, let alone Twitter. They spread through the exuberance of momentum and the rigid isolation of repressive rulers." Montefiore, "Every Revolution is Revolutionary in Its Own Way," *New York Times*, March 27, 2011.

[40] Stephen Williams, the BBC's executive editor for the Asia Pacific Region, emphasized the reciprocal relationship between online and television-based information flows during the 2009 Iranian protests. At the height of the protests, he notes, Internet photos, e-mails, and text messages arrived at BBC's Persian TV offices in London at the rate of between six and eight per minute. "The influence of Internet social media was huge in disseminating pictures and messages round the world," Williams says, and it "has undoubtedly helped the Opposition contact other like-minded voices inside Iran. But the most impact making pictures and reports could only be seen by a wider public in Iran through Persian-speaking TV, because Internet activity is limited." See Stephen Williams, "The Power of TV News: An Insider's Perspective on the Launch of BBC Persian TV in the Year of the Iranian Uprising," Joan Shorenstein Center on the Press, Politics and Public Policy Discussion Paper Series #D-54 (February 2010).

[41] See summary of an expert panel convened to address the role of social media in Middle Eastern political protests: Lauren Kirchner, "'Information Wars' on Al Jazeera English," *Columbia Journalism Review* (February 14, 2011).

[42] Jillian York, "More than Half a Billion Internet Users are being Filtered Worldwide," Open Net Initiative, January 19, 2010. Available at http://opennet.net/blog/2010/01/more-half-a-billion-internet-users-are-being-filtered-worldwide.

[43] Aday et. al., "Blogs and Bullets,"13.

[44] Clinton, "Internet Rights and Wrongs."

[45] Kathleen Hickey, "DARPA Looks for Stealthier Internet Access," *Defense Systems*, May 24, 2010. Available at http://defensesystems.com/articles/2010/05/21/darpa-safer-solicitation.aspx?admgarea=DS.

[46] "Tor: Sponsors" (no date), http://www.torproject.org/about/sponsors.html.en.

[47] David Talbot, "Dissent Made Safer," *Technology Review* (May/June 2009). Available at http://www.technologyreview.com/printer_friendly_article.aspx?id=22427. Also see question by Harvard Law School's John Palfrey to Ron Deibert, director of the University of Toronto's Citizen Lab: "What's going to happen when someone does something terrible using Psiphon, plans a terrorist attack, for instance? What's Psiphon's liability?" http://www.ethanzuckerman.com/blog/2007/01/31/ron-deibert-on-the-history-and-future-of-psiphon/.

[48] See, for example, Dan Froomkin, "Deciphering Encryption," *The Washington Post*, May 8, 1998, and, more recently, Dr. Marco Gercke, director of the Cybercrime Research Institute, "From Encryption to Failure of Traditional Investigation Instruments," writing on the Freedom from Fear blog, available at http://www.freedomfromfearmagazine.org/index.php?option=com_content&view=article&id=311:from-encryption-to-failure-of-traditional-investigation-instruments&catid=50:issue-7&Itemid=187.

[49] Jaikumar Vijayan, "U.S. Web Site Said to Offer Strengthened Encryption Tool for al-Qaeda Backers," *Computerworld*, January 23, 2008. Available at http://www.computerworld.com/s/article/9058619/U.S._Web_site_said_to_offer_strengthened_encryption_tool_for_al_Qaeda_backers?taxonomyId=16&intsrc=hm_topic.

[50] Ed Black, president and CEO of the Computer & Communications Industry Association, "Google, the Internet and China: A Nexus Between Human Rights and Trade?" (Testimony before the Congressional Executive Committee on China, Washington, D.C., March 24, 2010).

[51] Korea-U.S. Free Trade Agreement Article 15.8 "Cross Border Information Flows," signed June 1, 2007, http://www.ustr.gov/sites/default/files/uploads/agreements/fta/korus/asset_upload_file816_12714.pdf.

Part  5

CYBERSPACE: NEW POLICIES AND A
NEW STRATEGY

CHAPTER 9

A Path Forward for Cyber Defense and Security

Michael Chertoff
Co-Founder and Managing Principal
The Chertoff Group
Former U.S. Secretary of Homeland Security

John Michael McConnell
Vice Chairman
Booz Allen Hamilton
Former U.S. Director of National Intelligence

"Forcing cybersecurity into a simplified unitary framework limits our choices and underestimates the complexity of the most novel and serious disruptive threat to our national security since the onset of the nuclear age sixty years ago."

—MICHAEL CHERTOFF & JOHN MICHAEL MCCONNELL

A Path Forward for Cyber Defense and Security

Michael Chertoff
Co-Founder and Managing Principal
The Chertoff Group
Former U.S. Secretary of Homeland Security

John Michael McConnell
Vice Chairman
Booz Allen Hamilton
Former U.S. Director of National Intelligence

In 2008, President George W. Bush ordered the launch of the Comprehensive National Cybersecurity Initiative (CNCI), a now-declassified twelve point strategy to address cybersecurity threats across the civilian, military, government, and private domains. The Department of Defense and the Department of Homeland Security convened a group of government and business leaders to address cybersecurity issues, under the Enduring Security Framework. Shortly after taking office, President Barack Obama ordered a review of the CNCI and subsequently reaffirmed the mandate to proceed with a national cyber initiative. President Obama appointed a White House official to coordinate the strategy and Congress has taken up possible legislation.

Despite these various government initiatives, there is no comprehensive strategy for cyber defense and security in place. Last year, Deputy Secretary of Defense William Lynn described the Defense Department's evolving approach to defending against cyberattacks, which are escalating as a serious counterintelligence and warfighting issue. Soon thereafter, Deputy Secretary of Homeland Security Jane Lute responded with an opinion piece asserting that the Internet is not a war zone and arguing for a number of measures that the private sector can undertake to reduce its vulnerabilities to cyberattacks. This was followed by a DHS paper that elaborated on some characteristics of a more secure cyber "ecosystem." This past summer, the Department of State issued an international cyberstrategy and the Department of Defense announced a cybersecurity information-sharing pilot with several major defense companies. At the same time, the administration offered a legislative proposal to promote cybersecurity among operations of critical infrastructure.

But while these initiatives approach and characterize the challenge of threats to our cyber systems, they do not yet amount to a unified vision of the problem and solution sets. Indeed, it sometimes seems that those examining the problem are talking past each other. At one end of the spectrum are those who portray cyber risks as verging on the catastrophic, sketching cyber combat scenarios that result in extinguishing our civilization. At the other end of the spectrum are those who claim it is all overblown, and that the issue of cybersecurity is about updating virus protection and good police work. To those who have been around the security community over the last decade, this will sound much like the familiar debate about terrorism, between those who claim it is a criminal problem to be addressed by law enforcement and those who argue that terrorists have declared a war that must be fought with military capabilities.

In fact, the dichotomy between these approaches is oversimplified in the case of terrorism, and even more inadequate to define a strategy for protecting our cyber assets. Forcing cybersecurity into a simplified unitary framework limits our choices and underestimates the complexity of the most novel and serious disruptive threat to our national security since the onset of the nuclear age sixty years ago. Cyber threats will sometimes be a central dimension of military posturing and warfighting, and when they are critical they will require the response of all elements of national power. On the other hand, much destructive activity is occurring at the commercial and individual level, where military approaches are ill-suited and where the actors are largely within the private sector. If we debate the way forward in protecting cyber assets as a philosophical choice between "militarizing the Internet" or letting the market play the primary role, we rob ourselves of the full range of resources that we might mobilize.

Our ability to fully develop and implement national strategies for cybersecurity is also hampered by a tendency of government agencies to examine the problem from the perspective of their own authorities and capabilities. Abraham Maslow famously said that when you carry a hammer, everything looks like a nail; our agencies carry different tool sets and often view problems as whatever they can fix using the tools they carry. Intelligence agencies, in particular, are conditioned to sharply restrict their activities within the United States and as relating to U.S. persons—and rightly so. But while there are legal rules that require this, the non-constitutional limitations can—and should—be modified by lawmaking if there is good reason to do so. Likewise,

Congress can use legislation to affect the respective roles of the government and the private sector in incentivizing or driving certain forms of cyber behavior. In other words, our solutions to cyber threats should not be a function of what we think we can do with the rules and tools that we have; those rules and tools should be crafted based on the development of a cyber defense and security (CDS) doctrine that sets forth our strategic objectives and the roles and responsibilities of government and private institutions across all the domains touched by cyber activities.

How do we develop a comprehensive CDS doctrine? Doing so begins with an appreciation of the scope and nature of the threats. From that understanding, we should elaborate a doctrine that sets forth our national objectives in securing ourselves and allocates defense responsibilities between government and the private sector. The doctrine should also address allocation of government responsibilities among agencies, delineating which objectives each is responsible for achieving. A critical feature of developing this doctrine is balancing the various goals of security, privacy, freedom, and economic prosperity. With that framework set, Congress can enact or adjust the authorities appropriately to allow execution of the doctrine, subject to constitutional or civil liberties constraints. This article begins the process of posing questions that must be answered to develop the strategy under the preceding template.

Threats and consequences

While it is fair to say that the Internet is not a war zone, it could certainly become one. Moreover, war-like activity has been experienced as recently as 2007 and 2008. In 2007, Estonian government and financial institutions were the object of massive denial of service attacks aimed at disrupting and denying their ability to function. And when Russia invaded Georgia in 2008, ground movements were accompanied by cyberattacks aimed at disrupting Georgian command and control functions. Indeed, the United States-China Security Commission—a Congressionally mandated body— has identified cyberwarfare as an explicit part of Chinese military doctrine.

But most cyberattacks are not this dramatic nor so obviously tied to classically war related activities. Recent media reporting reveals intrusions into financial institutions such as Nasdaq, theft of data from energy companies, exfiltration of data from Google, massive identity thefts, and financial frauds. Much of this activity is directed by criminal groups, although nation-states also use the Internet for intelligence purposes. While these are not always physically destructive cyber activities, they can cause extremely serious personal and economic damage on a national scale. As

Deputy Secretary William Lynn's 2010 Foreign Affairs article has made clear, huge volumes of sensitive commercial information and intellectual property are stolen on a regular basis.[1] These data thefts directly affect our global competitiveness. Identity theft and credit fraud erode public trust in the Internet, which in turn negatively impacts investment and trade activity. On a personal level, there are heart-rending stories of personal financial and reputational trauma caused by organized cybercrime and thievery.

While all of these threats can have serious consequences, the responses to each may be different in scale and type, and the appropriate allocation of responsibility will vary. Accordingly, it is helpful to disaggregate the cyber threats which we face into several categories.

Data theft involves the unauthorized and often undiscovered exfiltration of confidential or proprietary data from a system. This may include intellectual property, business sensitive information, confidential government information, and classified national security information. *Fraud* involves using cyber tools to steal or deprive a victim of money, information, or property (including personal information) by deceiving the victim into paying the money or furnishing the property or information under false pretenses. *Denial of service attacks* interfere with access to or use of networks by overwhelming the network with data or commands so that its capacity to process additional data or commands is exceeded. This disrupts but does not necessarily damage or destroy the system under attack. *Destructive attacks* damage, destroy, or otherwise take control of a victim's computer systems. The consequences may range from denial of use to corruption or outright destruction of networks and systems, including those elements of physical infrastructure that are dependent on those systems.

Although popular culture reinforces the impression that the most significant threats are launched by attacks hacking into targeted systems, in fact, devastating attacks can originate from different vectors. To be sure, malware can be introduced over the network by remote hacking. But malware is often introduced through a corruption of the supply chain that embeds it within hardware or software. Equally dangerous are viruses that are introduced into a network by deceiving an authorized user into inviting them (for example, through phishing), or through accidental or intentional compromise by an insider.

1, there a need for UK Defence to develop a Cyber & SIGINT Defence & Security doctrine?

Foundations of a Cyber Defense and Security doctrine

What are the objectives of a CDS strategy? To establish a secure cyber environment within which public and private institutions can operate without excessive risk that systems will be crippled or damaged, or that valuable assets will be misappropriated or injured. But those ends coexist with other important objectives, such as fostering economic efficiency and creativity and protecting privacy and individual rights. The development of a strategy for securing cyberspace, therefore, must balance these objectives and consider the cost-effectiveness of various approaches. That amounts to cyber risk management.

From a defense and security standpoint, cyber risks differ from traditional security risks because of the degree to which they play out in the private sector. Traditional consequential defense and security responsibilities are largely exercised by public authorities, such as the military or police. While private institutions may equip themselves against relatively low-level security threats using private guards, locks, and alarm systems, modern civil society does not expect—or even accept—that the responsibility or authority for security against major physical threats should rest largely in private hands. No one suggests that civilian society equip itself with the responsibility to repel enemy invasions, and outside of private enclaves, we do not rely on private entities to police our streets.

What should the government's responsibilities and objectives be in the realm of cyber defense and security? Unlike the physical world, where major national security threats are largely—although not entirely—external, cyberattacks on privately owned networks might well be carried out—and even mounted—from or through platforms that are privately owned and domestic. A crippling of the power grid or our major financial institutions could have a catastrophic national impact, comparable to the effects of a major physical attack. But traditional perimeter military defenses would be irrelevant.

Some argue that cyber defense and security, therefore, is best left to the market and individual initiative and innovation. While it is true that the private sector has unleashed enormous creativity in developing aspects of our cyber economy, it is far from clear that market incentives will be sufficient to spur adequate investment in cybersecurity. Left to their own devices, few private companies would invest more in securing their cyber assets than the actual value of those assets. Yet in an interconnected and interdependent world, the failure of one part of the network can have devastating collateral and cascading effects across a wide range of physical,

[handwritten margin note at top: Remains incumbent upon the mil. to take appropriate the used and develop and resilience capabilities defence / resilience]

[handwritten margin note at left: Considerations for any future strategy]

economic, and social systems. Thus, the marketplace is likely to fail in allocating the correct amount of investment to manage risk across the breadth of the networks on which our society relies.

At one extreme, one could argue that the government should own a monopoly over cyber defense and security, assuming total responsibility for protecting public and private networks, and operating network defenses, accrediting hardware and software, and developing rules to reduce insider threats. At the other extreme, government would disclaim any responsibility in this sphere, leaving the market and individual initiative to address these problems. Both options are unrealistic.

Rather, in allocating responsibilities for CDS among government and private actors, we need to consider: (1) who owns the network, asset, or system we seek to protect; (2) how critical that network, asset, or system is to vital or critical national interests, especially the interests of collateral or third parties; (3) the nature and potential effects of the threat to be addressed; (4) whether government or private parties are best situated to respond quickly and effectively to the threat, given the architectural and economic features of the Internet; and (5) civil liberties and privacy constraints.

Naturally, the government's greatest role and responsibility will be directed at defense and government systems. These are owned by government agencies, and by definition most will be of national importance or at least networked to systems of national importance. As owner of military and civilian government systems, government is operationally and legally positioned to maintain awareness of what occurs in these systems and to protect them.

Responsibility should be shared—with a fair degree of government involvement—for those privately owned networks and systems which are deemed critical based on interdependency or the essential nature of the services provided. Ownership and control of these networks are in private hands, but the ramifications of security failure in critical networks have a much broader scope. Because the effect of intrusions into these critical systems can be magnified for interdependent third parties, market-based incentives alone may not be sufficient to drive enough investment in security. Government is therefore a particularly important partner because it can leverage what Deputy Secretary Lynn described as "government intelligence capabilities to provide highly specialized active defenses."[2]

But even if government is to be an active partner in managing cyber defense and security for privately held critical infrastructure, the specific methods and tools that government employs can still be sculpted to minimize intrusions on private economic concerns and civil liberties. The government can promote defense and security in several (overlapping) ways: (1) *Warning and situational awareness*; alerting potential targets about detected threats. One possibility is shared situational awareness through a common operating picture of the network. (2) *Defense*; actively blocking malware or other attack tools. (3) *Target hardening*; taking measures to make target networks and systems less vulnerable, such as by encrypting data; using hardware and software to promote better "cyber hygiene," including access controls, limits on downloading, internal network monitoring and tracking; and validating hardware and software from the supply chain. (4) *Investigation and forensics*; actions taken to discover penetrations that have already occurred and to investigate their source. Where practical and appropriate, this effort can include prosecution of those who have mounted the attack. (5) *Prevention*; preventing attacks before they are launched by incapacitating the attack vector or the individuals trying to mount the attack. Incapacitation can be accomplished using legal process, cyber means, or even physical means. (6) *Resilience*; building capabilities to survive and mitigate the effects of cyberattacks by creating redundancies, traffic management tools, and other mechanisms.

In the case of each of these approaches, the government can choose to execute the approach itself or to encourage, enable, and/or require the private sector to execute the approach. For example, government will want to maintain a monopoly of control over acts of prevention that involve incapacitating attackers operating from platforms or servers overseas. That means that government alone could exercise the legal authority to defend against persistent cyberattacks by attacking the offending platform either using cyber tools or even physical means.

By contrast, it is likely government would want to leave in private hands much of the responsibility for hardening or reducing vulnerabilities of private systems, albeit with encouragement and possibly assistance from the government. In those areas where the government is not likely to intervene directly, for example in building resilience across private networks, it could still deploy a variety of measures to prompt the private sector to execute defensive or security measures. These tools include (in increasing order of coerciveness): (1) providing actionable information and best practices; (2) creating legal incentives and immunities for private action (including liability protection); (3) monitoring and assisting in operating defenses upon invitation or consent; and (4) forcing action through regulatory mandates or disclosure obligations.

The more intrusive and coercive techniques for driving various security measures into the private sector are obviously more likely to clash with protection of private property and civil liberties. By the same token, less heavy-handed tools, such as information sharing and legal incentives and immunities, are far less likely to engage controversy and should be considered in the first instance in dealing with the kinds of threats—such as data theft or computer crime—that are relatively lower on the consequence scale. Promoting government engagement in these less controversial ways provides an early opportunity to manage down cyber risks, even as we debate the role of government in addressing more sophisticated and higher consequence cyber threats, such as national security espionage or sabotage of our cyber infrastructure.

Evolving a doctrine

The foregoing landscape of risks, capabilities, and public and private interests provides the canvas on which decision-makers must strike the balance between competing goals of security, efficiency, privacy, and free movement over the Internet. Where the government assumes responsibility for executing cybersecurity, doctrine refines specific policy principles.

For example, if the government exercises a monopoly over the right to prevent attacks by responding with force—using either cyber or physical tools—it must decide how and when it will trigger the response in connection with different types of threats. Acts of espionage or data theft—which are the modern analog to old-fashioned spying—may well be regarded as insufficient to trigger retaliatory or preemptive action, because the U.S. government has not generally treated espionage by foreign powers as in itself an act of war warranting forceful response. On the other hand, a foreign nation's attack on the integrity of important command and control systems or critical infrastructure may well be sufficiently consequential to warrant response in force. Indeed, as during the Cold War, one element of a response doctrine in such cases should be announcement of a declared policy of active prevention or retaliation under certain specified circumstances. Another important element of a response in force doctrine would be elaboration of the type and nature of evidence deemed sufficient to attribute an attack to a particular actor.

At the other end of the security spectrum, where government shares security responsibilities with the private sector, a doctrine will be necessary to clearly set forth the expectations of both the public and private sectors regarding their shared obligations. When the government chooses to enable private sector security measures by engaging in warning, the doctrine should set forth when, how, and with what degree of assurance warning will occur. A further decision will be whether the government should, by invitation, share tools for gaining situational awareness with operators of a private network.

When the government chooses to regulate, the doctrine should determine whether the regulation will be highly prescriptive or simply set objectives and broad metrics, leaving flexibility for implementation to the private sector. And where the government engages in active monitoring or defense, the doctrine should set forth how government agencies will treat and share information they obtain.

Finally, once a whole government doctrine is set, leaders should turn to the subsidiary issue of how to allocate the responsibilities that the government bears among various agencies, including intelligence agencies, law enforcement agencies, and regulators. All too often, evolution of government doctrine begins with agencies forging policies that are designed to expand or enhance their existing capabilities or authorities. But strategy should not be the handmaiden of interagency competition. Only when government roles, responsibilities, and functions have been formulated does it make sense to determine which organizations are best suited to execute them based on their intrinsic capabilities and statutory purposes.

Rewriting authorities

After the doctrine is designed, it must be matched against existing authorities to determine whether they need to be amended or new ones need to be created. The outer boundaries are, of course, set by the Constitution. Within those bounds, the doctrine should reflect privacy and other civil liberty concerns, and authorities can then be constructed to protect those concerns against encroachment. In dividing authorities among agencies, a balance should be struck between, on the one hand, assignment of authority to those who are best situated to discharge responsibility and, on the other, the desire to prevent undue concentration of power and to ensure institutional mechanisms to prevent abuse.

But authorities should not be drafted as a means to ring-fence bureaucratic turf against encroachment. Moreover, some long-held legal restraints on agency action will have to be revisited if government is to play a serious role in promoting cyber defense and security. For example, venerable and strongly held restrictions against intelligence agencies collecting information inside the United States or involving U.S. persons are difficult to apply when agencies are asked to participate in monitoring or defending global cyber networks that route packets through the United States. Should the monitor's ability to function depend on the happenstance of whether a hop point in the routing process is located on a U.S.-based server? Should the restriction be modified when the monitoring is not designed to collect the content of the cyber traffic, but simply to inspect individual packets to determine whether malicious code is embedded, or to watch traffic flow patterns to look for anomalies or suspect IP addresses?

If our strategy and doctrine conclude that the government should play a role in network monitoring and shared situational awareness—at least with the consent of network owners and operators—then it makes no sense to exclude or limit the authority of the appropriate intelligence agencies in that mission. In that way, the legal rules of the road are crafted to enable government to execute our national cyberstrategy, rather than subordinating the optimal strategy and doctrine to a set of legal rules largely built in a different era.

Michael Chertoff is Co-Founder and Managing Principal of The Chertoff Group, a global security advisory firm that provides consulting, business development and merger and acquisition (M&A) advisory services for clients in the security, defense and government services industries. During 2004 – 2009, Mr. Chertoff served as Secretary of the U.S. Department of Homeland Security, where he led the federal government's efforts to protect our nation from a wide range of security threats. Earlier in his career, Mr. Chertoff served as a federal judge on the U.S. Court of Appeals for the Third Circuit and head of the U.S. Department of Justice's Criminal Division where he investigated and prosecuted cases of political corruption, organized crime, corporate fraud and terrorism – including the investigation of the 9/11 terrorist attacks. Mr. Chertoff is a magna cum laude graduate of Harvard College (1975) and Harvard Law School (1978). In addition to his role at The Chertoff Group, Mr. Chertoff serves as senior of counsel at Covington & Burling LLP.

Mike McConnell is Vice Chairman of Booz Allen Hamilton, where his primary roles include serving on the firm's Leadership Team and leading Booz Allen's rapidly expanding cyber business. After retiring from the Navy in 1996 as a Vice Admiral, Mr. McConnell joined Booz Allen, and led the development of the firm's Information Assurance business and the firm's Intelligence business focused on policy, transformation, homeland security, and intelligence analytics, rising to the position of senior vice president. Upon being asked by President George W. Bush in 2007 to become the second Director of National Intelligence, he left Booz Allen and served as the DNI for two years under Presidents Bush and Obama. In 2009, Mr. McConnell returned to Booz Allen as an executive vice president to lead the firm's Intelligence business. In 2011 he was elevated to his current position as Vice Chairman of the firm. Mr. McConnell's career has spanned over 40 years focused on international development and foreign intelligence issues. His 29-year distinguished career as a U.S. Navy intelligence officer included several significant assignments. During Desert Shield/Storm and the dissolution of the Soviet Union, Mr. McConnell served as the Intelligence Officer for the Chairman of the Joint Chiefs of Staff, General Colin Powell, and the Secretary of Defense, Dick Cheney. From 1992 to 1996 he served as the Director of the National Security Agency (DIRNSA) under Presidents George H.W. Bush and William J. Clinton. Mr. McConnell holds an M.P.A. degree from George Washington University, is a graduate of the National Defense University (Global Telecom), the National Defense Intelligence College (Strategic Intelligence), and holds a B.A. degree in Economics from Furman University.

---

[1] William J. Lynn, "Defending a New Domain: The Pentagon's Cyberstrategy," *Foreign Affairs* (September/ October 2010).

[2] Ibid., 103.